p.154

BECKY A. SIGMON is Associate Professor in the Department of Anthropology and a member of the Department of Anatomy at the University of Toronto.
JEROME S. CYBULSKI is Research Scientist and Chairman of the Physical Anthropology Programme, National Museum of Man, National Museums of Canada, Ottawa.

Homo erectus, forerunner of *Homo sapiens* survived for more than one million years geographically dispersed over the Old World, ranging south into Africa, north into western Europe, and east into Indonesia and China. The species occupied a middle position in the evolutionary lineage of hominids, and thus is critical for understanding earlier and later developments in human paleontology.

This collection of fifteen essays refocuses attention on this major phase of human evolution and, simultaneously, honors the memory of Dr Davidson Black on the occasion of the fiftieth anniversary of his discovery of *Sinanthropus pekinensis* at Chou Kou Tien.

The articles, written by internationally recognized experts in the field of human paleontology, provide an updated account of discoveries and interpretations from historical, regional, and thematic perspectives.

Historical reviews of the life and work of Black provide biographical and bibliographical information as well as an account of the early discovery of Peking man and an assessment of the scientific value of the casts of the fossils made prior to their loss during the Second World War.

Subsequent chapters present theoretical approaches to the study of *Homo erectus* as known today, and the temporal, geographical, and morphological diversity of the fossils. These essays also describe recent research, some of it here in English for the first time, into variants of *Homo erectus*, and predecessors and descendants of this group, which have been discovered in Asia, Europe, and Africa from the 1920s to the present.

This book will be of interest to historians and philosophers of science as well as anatomists, paleoanthropologists, and human paleontologists.

北京人
懷念故步達生醫生論文集

BECKY A. SIGMON and
JEROME S. CYBULSKI
Editors

Homo erectus
Papers in Honor of
Davidson Black

Based on the proceedings of an international symposium in honor of
Davidson Black, Cedar Glen, Ontario, October 21, 1976

UNIVERSITY OF TORONTO PRESS
Toronto Buffalo London

©University of Toronto Press 1981
Toronto   Buffalo   London
Printed in Canada

ISBN 0-8020-5511-7

Canadian Cataloguing in Publication Data

Main entry under title:
Homo erectus
  (Symposia of the Canadian Association for Physical
  Anthropology ; v.1)
  Proceedings of a symposium held at Cedar Glen, Ont.,
  Oct. 21, 1976.
  Bibliography: p.
  Includes index.
  ISBN 0-8020-5511-7

1. Black, Davidson, 1884-1934 - Congresses. 2. Pithecanthropus erectus - Congresses. 3. Fossil man - Congresses. I. Sigmon, Becky A. (Becky Ann), 1941- II. Cybulski, Jerome S., 1942- III. Black, Davidson, 1884-1934. IV. Canadian Association for Physical Anthropology. V. Series.
GN282.H65   573.3   C81-094115-5

Case: *Sinanthropus pekinensis*. Top view is a reconstruction by Professor Franz Weidenreich. Bottom view represents Skull II from Locus L, Chou Kou Tien. (Photographs are of University of Pennsylvania Museum fossil reproductions.)

Photograph of Davidson Black from D. Hood, *Davidson Black, a biography* (Toronto: University of Toronto Press 1964).

Calligraphy by Miss Ka Bo Tsang.

Symposia of the Canadian Association for Physical Anthropology
Volume 1

Ashley & Crippen

DAVIDSON BLACK   1884–1934

Contents

FOREWORD ix
EDITORS' PREFACE xi
CONTRIBUTORS xiii

Introduction / Becky A. Sigmon 3

1 The works of Davidson Black / Jerome S. Cybulski and Paul Gallina 13

2 Davidson Black: an appreciation / Harry L. Shapiro 21

3 Davidson Black, Peking Man, and the Chinese Dragon /
   G.H.R. von Koenigswald 27

4 The Significance of the *Sinanthropus* casts and some paleodemographic
   notes / Alan Mann 41

5 *Homo erectus* in human descent: ideas and problems / W.W. Howells 63

6 Solo Man and Peking Man / Teuku Jacob 87

7 The position of the Vértesszöllös find in relation to *Homo erectus* /
   Andor Thoma 105

8 Les Anténéandertaliens en Europe / Marie-Antoinette de Lumley 115

9 *Homo erectus* in middle Europe: the discovery from Bilzingsleben /
   Dietrich Mania and Emanuel Vlček 133

10 Some views of *Homo erectus* with special reference to its occurrence in Europe / F. Clark Howell 153

11 Les hommes fossiles du Pléistocène moyen du Maghreb dans leur cadre géologique, chronologique, et paléoécologique / J.-J. Jaeger 159

12 *Homo erectus* at Olduvai Gorge, Tanzania / G.P. Rightmire 189

13 The Koobi Fora hominids and their bearing on the origins of the genus *Homo* / Alan Walker 193

14 The Kohl-Larsen Eyasi and Garusi hominid finds in Tanzania and their relation to *Homo erectus* / Reiner Protsch 217

Homo erectus: a synopsis, some new information, and a chronology / Jerome S. Cybulski 227

MAP SHOWING FOSSIL SITES 237
BIBLIOGRAPHY 241
INDEX 265

Foreword

Almost any Canadian who has a general interest in the history and origin of the human species will have heard of 'Peking Man.' Few, however, would know the pivotal role played by a Canadian, Dr Davidson Black, in the acquisition of knowledge about this member of the hominid family. Within the scientific community the association of Davidson Black with *Sinanthropus pekinensis* is almost equally rare. More widely known in this connection are the names of Franz Weidenreich, Pei Wen-Chung, and Père Teilhard de Chardin.

The neglect of Black by both scientists and laymen in Canada is in part due to his untimely death in 1934, just seven years after the first in situ discovery of *Sinanthropus* at Chou Kou Tien. Other factors that contributed to his anonymity among his countrymen are historical circumstances and, perhaps, the national temperament.

Davidson Black's research on *Sinanthropus pekinensis* spanned the period 1927 to 1934. Five of those years were depression years in Canada, and the times were not conducive to fostering public interest in prehistoric man. In Black's 1928 lecture to the Royal Canadian Institute on his views concerning hominid origins, the audience was unimpressed. Their skepticism may have arisen from seeing the evidence Black presented in support of his claims: a single molar tooth. Perhaps their reservation was based on a tendency to disregard a native son until he had proved his mettle elsewhere. Whatever the reasons, it remained for the anthropological community abroad to recognize Davidson Black's achievements before accolades were forthcoming from home. The sole such acknowledgment was the conferring (in absentia) of the honorary degree of Doctor of Science on Black in 1930 by the University of Toronto, his alma mater. Within four years of that occasion Davidson Black was dead. His friends and colleagues, W.C. Pei, Teilhard de Chardin, and C.C. Young, continued the work they had begun together; his successor, F. Weidenreich, completed the analysis of the hominid fossils that the Chou Kou Tien site ultimately yielded. In Canada, Davidson Black passed into history, recalled only in an occasional newspaper paragraph as 'the forgotten Canadian.'

In the fall of 1975, the Canadian Association for Physical Anthropology/ l'Association pour l'Anthropologie Physique du Canada decided that it was time to rectify past omissions. In association with the National Historic Sites and Monument Board of Canada, we wished to pay tribute to the accomplishments of Davidson Black in a way that was both scholarly and personal. We agreed to organize a symposium that would include biographical and historical sketches of the man, as well as provide a wider perspective that Black the scholar would have appreciated. Thus a discussion of *Homo erectus*, of which *Sinanthropus* is but a variant, was warranted.

The papers that are presented in this volume are the results of that resolution. The Symposium on *Homo erectus* in Honor of Davidson Black, at which the majority of these papers was given, was made possible by the joint efforts of many individuals. The core group included Franklin Auger, Jerome S. Cybulski, Hermann Helmuth, Christopher Meiklejohn, F. Jerome Melbye, Becky A. Sigmon, as well as myself.

There is little doubt that our tribute to Davidson Black would not have been possible without the generosity of several institutions. We are deeply grateful to the National Museum of Man, National Museums of Canada, the Ontario Heritage Foundation, and the University of Toronto for their financial support of the symposium, held at Cedar Glen, Ontario, on October 21, 1976.

EMÖKE J.E. SZATHMARY
President, Canadian Association
for Physical Anthropology

Editors' preface

The circumstances that led to the formulation of this volume are discussed in the Foreword and in the Introduction. The latter identifies the participants in the Symposium on *Homo erectus* in Honor of Davidson Black and those authors who did not participate but very kindly submitted manuscripts to round out the subject matter of the volume. We asked all participants to expand their oral presentations as they deemed necessary, and to incorporate insights gained during the meeting as well as information they might exchange among themselves after the symposium. Once the manuscripts were submitted, we offered suggestions for greater clarity of presentation, including stylistic conformities. One of the participants, F. Clark Howell, elected not to submit a formal manuscript; therefore, with his permission, we have included an edited transcript of his oral presentation.

All of the references cited in the articles are presented in one bibliography at the end of the volume. We felt that a single, encompassing bibliography would itself form a useful part of the volume.

As is always the case in this type of endeavor, many people contributed to the preparation of the manuscript. Inevitably, we would unconsciously omit someone if we tried to name them all. We very much appreciate the secretarial aid and moral support of Mrs Clara Stewart, Erindale College, University of Toronto, and the typing skills of the secretarial pool at the Archaeological Survey of Canada, National Museum of Man. We thank the staff of the National Museums of Canada Library for assistance in verifying bibliographic references, and Cathie Twiss, archivist with the Archaeological Survey of Canada, for her initial organization of the bibliography and for her invaluable help with proofreading typed copy. We are grateful to Hermann Helmuth of Trent University for suggesting the inclusion of the paper on Bilzingsleben and for his translation of the original German-language manuscript. E.J.E. Szathmary's contributions, especially in the organization of the symposium, are gratefully acknowledged. Finally, we extend our deep appreciation to the authors whose work formed the core of the symposium and the production of this volume.

The book has been published with the help of a grant from the Social Science Federation of Canada, using funds provided by the Social Sciences and Humanities Research Council of Canada, and a grant from the Publications Fund of the University of Toronto Press.

BECKY A. SIGMON & JEROME S. CYBULSKI

Contributors

JEROME S. CYBULSKI National Museum of Man, National Museums of Canada, Ottawa, Ontario, Canada K1A 0M8

PAUL GALLINA National Museums of Canada, Ottawa, Ontario, Canada K1A 0M8

F. CLARK HOWELL Department of Anthropology, University of California, Berkeley, California, USA 94720

W.W. HOWELLS Peabody Museum, Harvard University, 11 Divinity Avenue, Cambridge, Massachusetts, USA 02138

TEUKU JACOB Faculty of Medicine, Gadjah Mada University, Yogyakarta, Indonesia

J.-J. JAEGER Laboratoire de Paléontologie, Académie de Montpellier, Place Eugène-Batallion, 34060 Montpellier, Cedex France

G.H.R. VON KOENIGSWALD Division of Paleoanthropology, Senckenberg Museum, Frankfurt am Main, Federal Republic of Germany

MARIE-ANTOINETTE DE LUMLEY Laboratoire de Paléontologie Humaine et de Préhistoire, Université de Provence, Centre St Charles, 13331 Marseille, Cedex 3 France

DIETRICH MANIA Landesmuseum für Vorgeschichte, Halle, German Democratic Republic

ALAN MANN Department of Anthropology, University Museum (University of Pennsylvania), 33rd and Spruce Streets, Philadelphia, Pennsylvania, USA 19174

REINER PROTSCH Department of Paleoanthropology and Archaeometry, J.W. Goethe Universität, Frankfurt am Main, Federal Republic of Germany

G.P. RIGHTMIRE Department of Anthropology, State University of New York, Binghamton, New York, USA 13901

HARRY L. SHAPIRO Department of Anthropology, American Museum of Natural History, Central Park West at 79th Street, New York, New York, USA 10024.
BECKY A. SIGMON Department of Anthropology, Erindale College, University of Toronto, Toronto, Ontario, Canada M5S 1A1
ANDOR THOMA Laboratoire de Paléontologie des Vertébrés et de Paléontologie Humaine, Université Catholique de Louvain, 3 Place Louis Pasteur, 1348 Louvain-la-Neuve, Belgium
EMANUEL VLČEK National Museum, Václavské námesti, Prague, Czechoslovakia
ALAN WALKER Department of Cell Biology and Anatomy, Johns Hopkins University, School of Medicine, Baltimore, Maryland, USA 21205

HOMO ERECTUS: PAPERS IN HONOR OF DAVIDSON BLACK

BECKY A. SIGMON

Introduction

Following the investigations of Davidson Black, Franz Weidenreich, and others in the 1920s through 1940s, *Homo erectus*, the traditionally viewed 'middle stage' of human evolution, has taken almost a back seat to the more recent, widely publicized discoveries and studies of earlier forms of Man, near-Man, pre-Man, and the Anthropoidea in general. The present collection of papers is an attempt to refocus on *Homo erectus* and to provide an updated account of discoveries and interpretations of this ancient and equally important link in the understanding of our evolution.

The volume is an outgrowth of an international symposium in honor of Davidson Black, the Canadian anatomist who identified and introduced the fossil form, *Sinanthropus pekinensis,* in 1927. This fossil population encompasses a singular group of Middle Pleistocene specimens in China that have been instrumental in defining the morphology and culture of *Homo erectus.* The symposium, held in Cedar Glen, Ontario, in October 1976, was organized by the Canadian Association for Physical Anthropology to coincide with a recognition of Black's scientific achievements by the National Historic Sites and Monument Board of Canada. A commemorative plaque was unveiled by the board at the Medical Sciences Building of the University of Toronto, Black's alma mater.

Publication of the symposium proceedings was planned in advance. It was our goal to bring together researchers whose contributions, when assembled in one volume, would present what we felt to be a much needed synthesis of the role of *Homo erectus* in human evolution. We had hoped to include articles from paleoanthropologists who had discovered or investigated *H. erectus* from every part of the Old World where it has been found, or thought, to occur. An obvious omission in this endeavor is a spokesman from and for the People's Republic of China. An invitation was extended to Woo Ju-Kang and other Chinese paleoanthropologists, but circumstances did not permit their participation. A second geographical area not treated in detail is South Africa. John T. Robinson, well known for his work in this area, was invited but unfortunately was not able to attend the symposium or contribute a paper to the volume. Notwithstanding these omissions, we feel that the book provides a reasonably extensive coverage of *Homo erectus*. Discoveries in South Africa and recent findings in the People's Republic of China have been considered in the presentation by W.W. Howells, and in the concluding summary chapter.

The authors who actually participated in the symposium include Harry L. Shapiro, Alan Mann, W.W. Howells, Teuku Jacob, Andor Thoma, Marie-Antoinette de Lumley, F. Clark Howell, Alan Walker, Reiner Protsch, and Milford Wolpoff.[1] The editors served as chairpersons. G.H.R. von Koenigswald and J.-J. Jaeger were unable to attend but kindly contributed papers for publication. We solicited contributions from Phillip Rightmire and from Dietrich Mania and Emanuel Vlček specifically for the volume in order that their important recent research on materials from Olduvai Gorge and Bilzingsleben might be included. In the past three decades, the research of all these people, as well as a host of others whose work is cited in the bibliography, has substantially added to our present understanding of the phase of human evolution during the time of *Homo erectus*.

There is one contributor to the symposium who is not represented by an article in the volume. This is Davidson Black III, a practicing Toronto eye surgeon. Having inherited his father's motion picture films on the Chou Kou Tien excavations and on the preparation of the *Sinanthropus* fossils, he kindly and enthusiastically showed the films during the symposium. In addition, he gave permission for the Department of Anthropology at the University of Toronto to copy the films so that this record of his father's paleontological research would be permanently preserved at the University. We are grateful to Dr Black and his family for their contribution to the success of the symposium, and for their cooperation in preserving paleontological materials.

The book is organized within three frames of reference: historical, regional, and thematic. It begins with a thoroughly researched account of Davidson Black's written contributions to anatomy, physical anthropology, and paleontology, including his publications on *Sinanthropus pekinensis*. As a result of the efforts of Jerome Cybulski and Paul Gallina, both the National Museum of Man in Ottawa and the Department of Anthropology at the University of Toronto

---

1 We regret that there is no paper representing Wolpoff's contribution to the symposium. Because of the unfortunate timing of his most recent research and the publication deadlines, he was not able to make the desired revisions on the manuscript that he had submitted and subsequently withdrew.

will have on file the complete works of the scientist whose achievements we honor with this volume.

Harry Shapiro's aptly titled biographical sketch of Davidson Black is, indeed, an appreciation of the man and his efforts toward furthering the study of human evolution. Through another eminent anthropologist's words, we see an outstanding personality. Shapiro leads us through the growth and development of an individual whose intellectual pursuits have left an indelible imprint on the science of paleontology.

The story of *Homo erectus* would be incomplete without the inclusion of a chapter by Ralph von Koenigswald. As a scholar of immense stature in the field of evolutionary studies, von Koenigswald has contributed numerous chapters to the story of human evolution. In addition to his scientific treatises in paleontology, he is particularly well known to students of the discipline for his foresight in hiding, and consequently saving, certain fossil material prior to the Japanese invasion of Indonesia during World War II. In his own words: 'For some time, my collections of Javanese and Chinese human teeth (among them the teeth of *Gigantopithecus*) were buried in milk bottles in our garden' (von Koenigswald, personal communication 1978).

The enthusiasm and innate curiosity of von Koenigswald and, especially in this case, Mrs von Koenigswald, are traits that must have featured prominently in their 'drugstore searches' for fossil teeth in China and Java. These delightful, if unorthodox paleontological methods resulted in the discovery of one new hominid taxon, *Gigantopithecus blacki*, the suggestion of the presence in Asia of an early hominid similar to *Paranthropus* of Africa (von Koenigswald's *Hemanthropus*), and additional *Homo erectus* teeth. Among other lessons, von Koenigswald has taught us that there is more than one way to stalk a hominid.

The fact that alternative routes are available for the study of human paleontology is reemphasized by Alan Mann. His research is being undertaken at a time when paleontologists raise their brows in disapproving furrows when they read of studies being undertaken on casts of fossil hominid specimens. The general feeling is that casts should not be used as resource material for a scientific paper. Nevertheless, replicas are useful for comparisons among fossils when it is not possible to assemble all the relevant original specimens in one location. Casts are particularly useful for assessing the possible affinities of newly found specimens during fieldwork and for preliminary studies in the laboratory.

Mann's study represents a unique set of circumstances. The original *Sinanthropus* fossils of the 1920s and 1930s, as far as we know, were irretrievably lost during World War II. Primary casts have been preserved in the American Museum of Natural History and in a few other institutions. Mann raises the problem of the reliability of these copies. How closely do they duplicate the original specimens from Chou Kou Tien? He has measured the primary casts and compared his results with those provided by Black and Weidenreich from the original fossils. With few exceptions, Mann demonstrates that his measurements on the primary casts reveal a high degree of correspondence with those made on the original fossils from the Chou Kou Tien Locality 1 sample. This finding should encourage others to complete studies on these detailed and accurately cast replicas for new interpretations regarding the *Sinanthropus* fossils.

Mann's examination of the casts reveals not only interesting demographic data, but his studies also suggest that there is more morphological variation in the Chou Kou Tien Locality 1 sample than had previously been indicated by the studies of Black and Weindenreich on the original material.

W.W. Howells' paper provides an orientation framework for the subsequent, regionally specific discussions of *Homo erectus*. Traditionally, Howells is known for his ability to unravel the loose ends of a subject and reconstruct the fragments into a comprehensible form. In his article, Howells has assembled scattered information and reformed it into a lucid review of the fossil evidence and the theoretical constructs that have had—and should have—bearing on our interpretations of what constitutes *Homo erectus*. He reminds us that a grasp of the concept and taxon is not to be had by viewing it as a static, catchall unit into which fossils can neatly be placed. Rather, we must think of *H. erectus* as comprising populations that had their own pasts, presents, and futures. His theme, to explore the 'meaning and usefulness of *H. erectus* now as a taxon and concept' is oriented toward examining the question of its origins, the amount of variation present during its evolutionary span, and its subsequent evolution into *H. sapiens*.

Howells' thematic introduction to *H. erectus* is followed by regional discussions of the taxon, its forebears, and its successors. The areas treated in detail are Indonesia, East, West, and middle Europe, and North and East Africa.

Indonesia, specifically the island of Java, is where the first fossil evidence for *Homo erectus*, Eugene Dubois' 1891 discovery of *Pithecanthropus erectus*, was found. Paleontological research in Java following World War II has been largely under the direction of Teuku Jacob. Unlike many of his colleagues, Jacob prefers to divide the early Indonesian hominines into three species, *Pithecanthropus modjokertensis*, *P. erectus*, and *P. soloensis*. These three taxa are dated respectively at 1.9 million, 830,000, and 300,000 years BP, with some suggestion of overlap between the latter two. Jacob emphasizes that the specific and subspecific statuses of these groupings are far from clear, but nonetheless presents a persuasive, morphologically based argument for their taxonomic separation. He suggests that comparisons should be made separately between each Indonesian species and other Asiatic hominids who lived during the same periods. His comparisons of *P. erectus* and *P. soloensis* with *P. (=Sinanthropus) pekinensis* indicate greater morphological affinity between either of the latter Javanese taxa and the Chinese taxon than between the two Javanese taxa themselves. Similar comparisons should be made, Jacob suggests, between the Chinese *P. (=Sinanthropus) lantianensis* and the two earlier Javanese forms, *P. erectus* and *P. modjokertensis*.

As Jacob points out, one of the urgent needs in Javanese paleontology is the gathering of more absolute dates. Additional information on chronological sequences in Indonesia would serve to strengthen our knowledge of the existing data and to clarify our understanding of hominid migration patterns in the Old World. Jacob's concluding reflections on cannibalism, a practice that some scholars attribute to *Homo erectus*, remind us of the difficulties that one encounters (and the caution that should be used) in attempting to reconstruct behavior from bones.

Reconstructing phylogenies from fossilized bones is also no easy matter, as can be seen from Andor Thoma's struggle to make good taxonomic sense of some east European specimens. *Homo (erectus seu sapiens) palaeohungaricus* is a taxonomic designation that Thoma has proposed for the remains of two individuals found in 1965 at the Mindelian occupation site of Vértesszöllös, Hungary. He suggests that the remains are a sample of a population that migrated from Asia Minor over a land bridge to the Balkans. The most likely ancestral forms, he feels, would be the Chinese and Javanese *Homo erectus* populations. Other fossil forms that might be the result of similar migration patterns are those from Petralona, Bilzingsleben, and Mauer.

The Hungarian fossils discussed by Thoma exemplify the point made by Howells, that is, that how we define the species, *H. erectus* and *H. sapiens*, affects the way we think of transitional hominid forms. The Vértesszöllös specimens appear to be morphologically intermediate, a situation accounted for by Thoma's earlier taxonomic designation. However, he emphasizes that in phylogenetic systematics they represent *H. sapiens*.

For Europe proper, Marie-Antoinette de Lumley has taken us through time from 800,000 to 120,000 years ago with a discussion of fossil hominines known from that time period. Although culturally modified implements are known in Europe from well over a million and a half years ago, human skeletal remains (a tooth and a jaw) are dated at no earlier than 800,000 years BP. From somewhat later times, during the early part of the Middle Pleistocene (500,000 to 300,000 BP), there are a few more hominid specimens that have been recovered from four different sites.

8 *Homo erectus*

By middle Mid-Pleistocene times, 300,000 to 120,000 BP, relatively larger samples of early man are known throughout Europe. Although a certain amount of polymorphism is present, de Lumley interprets these hominids to represent groups of relatively homogeneous populations that did not have the same morphological patterns as their African ancestors. It appears, she suggests, that after the hominids left Africa and settled in Europe, different evolutionary pressures were brought to bear on the populations. These new selective pressures led to the development of the morphological patterns that de Lumley refers to as 'Anténéandertalien.'

Forming a somewhat similar conclusion, Dietrich Mania and Emmanuel Vlček report that the hominid remains of Bilzingsleben are sufficiently different from other *Homo erectus* material to warrant their subspecific nomen, *H. erectus bilzingslebensis*. The site of Bilzingsleben in the German Democratic Republic has yielded rich faunal and artifactual material in addition to hominid fossils. Amino-acid racemization has produced a date of 230,000 years, and the fossil bed provides other information that the site can be dated to the Elster/Mindel – Saale/Riss interglacial period, or the Holsteinian of the north European glaciation scheme. The site is of particular value because of the information it has provided on the paleoecology and behavior patterns of *H. erectus*. The climate of the time was warmer than it is today, and the flora included deciduous mixed forests and shrubs. The people had a knowledge of the use of fire, and were both hunters and gatherers.

Clark Howell's paper offers an overview of the problems that research in Europe has instigated but has not solved. It is an edited transcript of his symposium presentation. Given the customary definition of the taxon, Howell states that *Homo erectus* did not exist in Europe. Instead, he acknowledges the 'Anteneanderthalians' introduced by de Lumley. He raises the following questions and suggests that further investigation will be needed before they can be answered. Which hominid stock, in terms of its morphological development, gave rise to the European 'Anteneanderthalians'? From what part(s) of the Old World did the ancestors of the European hominines come? When did the first people arrive in Europe? How much geographical variation exists among the representatives of the earliest European populations?

Howell reiterates W.W. Howells' emphasis on the importance of the body of already existing data. These data deserve to be studied for what they can tell us about a specific phase of human evolution that consisted of beings 'successful in their own right.' Additional studies on material that has already been accumulated should begin to provide some of the answers to the questions he has posed.

The continent of Africa has provided a veritable warehouse of fossil evidence for the emergence of the genus *Homo*. Northwest Africa, by virtue of its proximity, would appear to be a pivotal area for studies concerning the peopling of Europe. However, as Jean-Jacques Jaeger points out, there is no paleontological evidence to substantiate the possibility of hominid migration between northwest Africa and Spain. His thorough review of the climatic, geological, and faunal events in northwest Africa during the Middle Pleistocene indicates that animal migration occurred only in an east-west direction.

With respect to the overall evidence for the presence of hominids, the situation in North Africa parallels that of Europe. Skeletal remains are not known prior to 700,000 BP but many cultural artifacts have been found from earlier periods. In northwest Africa, the fossils of 700,000 to 200,000 BP show a general morphology suggestive of local adaptations that persisted throughout the period. Yet, Jaeger identifies internal variability that points to ancestral-descendant relationships between the earliest representative, that of Ternifine, and later populations like that of Salé. The characteristics of all of the Maghreb fossils suggest that they could be descendants of *H. erectus* and, probably, ancestors of *H. sapiens* who inhabit the area today. Whether one refers to these Middle Pleistocene fossils as late remnant *H. erectus* populations, as Jaeger suggests, or as pre-*H. sapiens* is, perhaps, irrelevant. That they fill yet another gap in the evolution of the *Homo* lineage from one more part of the Old World is a general point of agreement.

Another area of Africa that has produced fossil evidence for *Homo erectus* is the Rift Valley, including Olduvai Gorge in Tanzania and East Turkana in Kenya. In the summer of 1977, Mary Leakey invited Phillip Rightmire to study and write up the formal descriptions of the *Homo erectus* specimens, OH 9 and OH 12, from Olduvai Gorge. Shortly thereafter we invited him to contribute a preliminary report for the volume. Rightmire's study is important because it demonstrates the variation that may occur in members of the same taxon. Olduvai Hominid 9 is shown to have a much more robust cranium than Olduvai Hominid 12. The fossils represent two populations that were separated by about 300,000 to 500,000 years. Although the time gap may be cited to account for the variation, it is important to identify the evolutionary processes responsible. We look forward to Rightmire's final report on the study of those two important East African hominids [see p. 230 of this volume].

Alan Walker has added a dynamic flair to the field of human evolutionary studies. He has tossed around ideas with many of us in this field and his receptivity to new approaches has resulted in the structuring of some of these mental musings into testable hypotheses. In the paper presented here, Walker strives to view the data as dynamic and to see fossilized specimens as representatives of individuals who lived within a demographic and biotic setting. From this point of view, he discusses procedures for dealing with small samples of the fossil record that show large amounts of variability. A particularly vexing problem that the discoveries at East Turkana have uncovered is the occurrence of three morphologically distinct hominids at the same time horizon. Among them is the specimen, KNM-ER 3733, regarded by many as a 'look-alike' of Peking Man, yet about a million years older. Walker takes us through various combinations of possibilities that might explain the East Turkana data and then discusses the reasons for selecting the interpretation that he considers to be most likely.

A final problem to which Walker directs our attention is that of the dating of the KBS tuff, a volcanic ash layer that lies directly above the deposits that contained fossils like KNM-ER 1470, an early but surprisingly advanced hominine. Reaching agreement on the age of the now questionably dated tuff will, he suggests, affect our interpretation of the nature and speed of the adaptive radiation that ultimately lead to the grade of organization that characterizes *Homo erectus* [see pp. 239-30 of this volume].

In another part of the Rift Valley, just south of Olduvai, there are two sites, Eyasi and Garusi, that have produced early hominid fossils. Reiner Protsch describes the history of the finds and how they have been interpreted since their discovery by the Kohl-Larsen expedition to Tanzania in 1934-40. Early studies resulted in their being thought of as members of the taxon *Homo erectus*. Protsch's studies now indicate that the two sites are separated by more than three million years and that the fossils they have yielded are morphologically distinct. Eyasi I, consisting of fragmentary cranial parts, has been shown to belong, both morphologically and chronologically, to the taxon *H. sapiens rhodesiensis*. It has been amino-acid dated to 34,000-35,000 years BP. The Garusi hominids, consisting of a maxilla and teeth, appear to be temporally commensurate with the jaws and teeth recently recovered from the Laetolil Beds by M. Leakey. Protsch suggests that the Garusi specimens are *Homo* and probably ancestral to *H. erectus*.

Since Davidson Black's introduction of *Sinanthropus pekinensis* in 1927, the fossil evidence for Middle Pleistocene hominids in the Old World has expanded, as has our idea of the complexity of this time period. It is no longer sufficient to regard *H. erectus* as simply a 'middle stage' in human evolution. The contributions in this volume clearly show that we are dealing with a group of beings that successfully survived for more than a million years during which time they expanded into many areas of the Old World and developed a large amount of morphological variation.

Although many would regard this group as a single species, the morphological variation and geographic diversity have led to problems of nomenclature. The authors herein find varying solutions as they struggle with the problem of classifying hominines of the Middle Pleistocene. Jacob prefers to retain the generic name *Pithecanthropus* in reference to the Asian specimens. Thoma's later transitional Vértesszöllös hominines are designated *H. (erectus seu sapiens) palaeohungaricus*. De Lumley refers to the European hominines of 800,000 to 120,000 years BP as Anténéandertaliens. Mania and Vlček call the Bilzingsleben hominines a separate European subspecies, *H. erectus bilzingslebensis*. Jaeger prefers to use the binomen *H. erectus* to refer to all the Northwest African specimens of 700,000 to 200,000 years BP. The East African *H. erectus* fossils from Olduvai and East Turkana are referred to by Rightmire and Walker as *H. erectus* and *H. erectus*-like.

To some extent the problems of nomenclature are a result of our tendency to regard *H. erectus* as a static paleospecies. The more we know of this group, however, the more we become aware of its dynamic nature, represented by populations of hominines spreading throughout the Old World, adapting and evolving. Under such circumstances we would expect to see fairly extensive morphological differences in the overall population through time and space, a situation that has led to vagueness in the definition of other paleospecies as well. Taxonomic problems are compounded when one notes that the successive *H. sapiens* is also inadequately defined.

If we are to understand the alpha and omega of *H. erectus*, as well as the nature of the group during its evolutionary peak, it is crucial that we investigate the bases underlying the morphological variation present. Just as *H. erectus* has traditionally been used as a springboard from which to pursue studies of earlier and later forms of hominids, the editors suggest that this volume might be used as a springboard from which to further pursue questions concerning the 'stage' of human evolution represented by *H. erectus*.

12 Homo erectus

Davidson Black family at the unveiling of the commemorative plaque honoring Davidson Black II. From left to right are Adrienne, Davidson IV, Davidson III, and Lynne Black. The plaque reads:

DAVIDSON BLACK
1884-1934

Davidson Black was born and educated in Toronto. He had begun a career in medicine when Sir Grafton Elliot Smith interested him in the problem of fossil man. After World War I, Black accepted a post at the Pekin Union Medical College, considering China to be a likely field for his studies. There, in 1927, on the basis of a fossil tooth found at Chou Kou Tien, he identified a new genus and species of hominid, *Sinanthropus pekinensis*. This discovery of "Peking man" was subsequently confirmed by the excavations of W.C. Pei and a team of Chinese and European scientists working with Black. He died in China.

Né et éduqué à Toronto, Davidson Black avait embrassé la carrière médicale quand sir Grafton Elliot Smith éveilla sa curiosité pour les fossils humains. Après la Première guerre mondiale, il accepta un poste au Pekin Union Medical College, considérant la Chine comme un champ de recherche idéal. En 1927, à partir d'une dent fossilée trouvé à Chou Kou Tien, il identifia une nouvelle espèce d'hominidé, le *Sinanthropus pekinensis*. Cette découverte fut subséquemment confirmée par les fouilles de W.C. Pei et d'une équipe de savants chinois et européens, qui travaillaient avec lui. Black mourut en Chine.

National Historic Sites and Monuments Board of Canada
Commission des lieux et monuments historique du Canada

(Photograph courtesy of Department of Information Services, University of Toronto, 45 Willcocks Street, Toronto, Ontario M5S 1A1)

1

JEROME S. CYBULSKI & PAUL GALLINA

The Works of Davidson Black

Au cours d'une brève carrière professionnelle qui a duré de 1909 à 1934, Davidson Black a publié un total de 51 ouvrages. Les recherches que, jeune homme, il avait faites dans le domaine de l'anatomie, et notamment de la neuroanatomie, servirent de préparation aux recherches poussées qu'il entreprit plus tard sur *Sinanthropus pekinensis*. Au nombre de ses écrits se trouvent 12 articles importants centrés sur l'anatomie endocrânienne et sur des études phylogénétiques à partir de cinq classes de vertébrés. La diversité de ses intérêts dans le domaine de l'anthropologie physique nous apparaît dans des écrits sur l'anthropométrie et sur les restes osseux des habitants de la Chine préhistorique. En 1925, Black présenta une étude théorique sur la dispersion des primates en Asie et ses conséquences sur le lignage de l'homme. Entre 1926 et 1934, il publia 20 écrits sur Chou-kou-tien et sur *Sinanthropus*, au nombre desquels il faut inclure ses célèbres monographies parues dans *Palaeontologia Sinica* et le texte de la conférence Croonian qu'il prononça devant la Royal Society de Londres en 1932.

With this volume we specifically honor Davidson Black for his identification and introduction of the fossil hominid taxon, *Sinanthropus pekinensis*. As Dr Shapiro and other contributors point out, that pronouncement was based on a single tooth excavated from the now famous 'Peking Man' site at Chou Kou Tien in 1927. However, many more fossil remains were discovered at the site during the next ten years, discoveries that compounded and secured Davidson Black's recognition of this new variety of fossil man.

Davidson Black's short life (1984-1934) was remarkably full, as attested by his biographer, Dora Hood. In addition to the information in her book (Hood 1964), details of his life and works may be found in obituaries written by his close friends and associates, George B. Barbour (1935) and Grafton Elliot Smith (1934a, 1934b), in Dr Shapiro's (1974) book on Peking Man and in his contribution to the present volume, in J. Gunnar Andersson's (1934) *Children of the Yellow Earth*, and in a special memorial volume[1] which was published in Peking shortly after Dr Black's passing.

Dr Black's identification of *Sinanthropus* came eight years after his arrival at the Peking Union Medical College to take up an appointment delayed almost a year by his World War I service with the Canadian Army Medical Corps in Canada and in England. The college had been officially instituted by the New York-based China Medical Board of the Rockefeller Foundation in 1915 to promote the teaching and practice of modern medicine in China. Dr Black's appointment as professor of neurology and embryology was at the urging of Dr Edmund Vincent Cowdry, a friend of his student days, who had been appointed head of the anatomy department in 1918. Dr Black assumed Dr Cowdry's position upon the latter's resignation in 1921. In 1929, Black and Dr V.K. Ting became honorary co-directors of the Cenozoic Research Laboratory at the college.

Prior to his military service, Dr Black taught at Western Reserve University in Cleveland, Ohio, between 1909 and 1917. He had first been assigned a lectureship in the Department of Anatomy and was then appointed an assistant professor in 1913. He received a Bachelor of Medicine degree from the University of Toronto in 1906 and took a Bachelor of Arts Degree in 1909. In 1914 Dr Black studied with Grafton Elliot Smith in England.

Dr Black's research in anatomy, particularly neuroanatomy, was extensive. From 1913 to 1922 he published 12 major articles on the subject, most of which appeared in the *Journal of Comparative Neurology*, and read several papers at scientific meetings such as those of the American Association of Anatomists. It is significant that the bulk of this work centered on endocranial anatomy and on phylogenetic studies, the latter covering five classes of vertebrates. The knowledge gained from this research served him well in his later work with the *Sinanthropus* fossils.

Most of Dr Black's remaining published works were concerned with *Sinanthropus*. However, his wider interests in physical anthropology are reflected in a short article on anthropometry published in 1920 and in studies of the recent prehistoric people of China. The latter appeared between 1925 and 1928 and included two monographs published in the *Palaeontologia Sinica*.

---

1 *Davidson Black 1884–1934; in memoriam*, published by the Geological Society of China and Peking Society of Natural History in 1934. The volume contains tributes delivered at a memorial meeting in Peiping on May 11 by Dr Black's friends and colleagues: Dr Wong Wen Hao, Dr Paul H. Stevenson, Dr V.K. Ting, Dr Amadeus W. Grabau, Père Teilhard de Chardin, Dr C.C. Young, Dr George B. Barbour, Mr W.C. Pei, and Mr Roger S. Greene.

Just before his work with *Sinanthropus*, Dr Black contributed a major theoretical 'position' paper on primate radiation in Asia and its bearing on the ancestry of man. His publications on Chou Kou Tien and *Sinanthropus*, between 1926 and 1934, numbered 20. Major works were two monographs in the *Palaeontologia Sinica*, a coauthored memoir published by the Geological Survey of China, and publication of the coveted Croonian lecture he delivered to the Royal Society of London in 1932. This last was a crowning synopsis of his deep commitment to the study of *Sinanthropus pekinensis*.

Following is what we believe to be a comprehensive bibliography of Davidson Black's writings,[2] arranged topically:

ANATOMY–NEUROANATOMY

1913 The central nervous system in a case of cyclopia in *Homo*. *Journal of Comparative Neurology* 23 (3): 193-257
___ The study of an atypical cerebral cortex. *Journal of Comparative Neurology* 23 (5): 351-69
___ On the possible trophic role of the afferent projection fibers in the development of the cortex cerebri. Paper read before the Experimental Section of the Academy of Medicine, Cleveland, Ohio, Sept. 5, 1913. 4 pp. in *Black: Collected Papers. Volume II*. Library of the American Museum of Natural History, Osborn Library of Vertebrate Palaeontology
1914 Two cases of cardiac malformation – more especially of the infundibular region. *Journal of Anatomy and Physiology* 48: 274-9
___ The relation of the accessory nerve to the vagus complex. *Cleveland Medical Journal* 13 (2): 128-9 (abstract)
___ On the so-called 'bulbar' portion of the accessory nerve. *Anatomical Record* 8 (2): 110-12 (abstract) of the American Association of Anatomists (30th session)
1915 Brain in primitive man. *Cleveland Medical Journal* 14: 177-85
___ A note on the *sulcus lunatus* in man. *Journal of Comparative Neurology* 25 (2): 129-34
___ A study of the endocranial casts of Ocapia, Giraffa and Samotherium, with special reference to the convolutional pattern in the family of Giraffidae. *Journal of Comparative Neurology* 25 (4): 329-60

---

2 We thank the staffs of the Osborn Library of Vertebrate Palaeontology at the American Museum of Natural History and of the Library of the University of California at Berkeley for their help with our research. The research was supported by the National Museums of Canada Library and by the National Museum of Man.

____ Notes on the endocranial casts of Okapia, Giraffa and Samotherium. *Anatomical Record* 9 (1): 56-9 (abstract) of the American Association of Anatomists (31st session)

1916 Cerebellar localization in the light of recent research. *Journal of Laboratory and Clinical Medicine* 1 (7): 467-75

____ Endocranial markings of the human occipital bone and their relations to the adjacent parts of the brain, with special reference to the so-called 'vermiform fossa.' *Anatomical Record* 10 (3): 182-5 (abstract) of the American Association of Anatomists (32nd session)

1917 The motor nuclei of the cerebral nerves in phylogeny: a study of the phenomena of neurobiotaxis. Part I. Cyclostomi and Pisces. *Journal of Comparative Neurology* 27 (4): 467-564

____ The motor nuclei of the cerebral nerves in phylogeny. A study of the phenomena of neurobiotaxis. II. Amphibia. *Journal of Comparative Neurology* 28 (2): 379-427

1920 The motor nuclei of the cerebral nerves in phylogeny — a study of the phenomena of neurobiotaxis. III. Reptilia. *Journal of Comparative Neurology* 32 (1): 61-98

____ Studies on endocranial anatomy. II. On the endocranial anatomy of Oreodon (Merycoidodon). *Journal of Comparative Neurology* 32 (3): 271-327

____ Preliminary report on the endocranial anatomy of Oreodon. *China Medical Journal* (Anatomical Supplement) 34 (4): 19-20 (abstract)

____ The significance of certain endocranial markings in man and the importance of endocranial anatomy from the standpoint of anthropology. *China Medical Journal* 34 (6): 689-90 (abstract). Proceedings of the Anatomical and Anthropological Association of China

1922 The motor nuclei of the cerebral nerves in phylogeny. A study of the phenomena of neurobiotaxis. IV. Aves. *Journal of Comparative Neurology* 34 (2): 233-75

ANTHROPOMETRY

1920 Concerning anthropometry and observations on healthy subjects. *China Medical Journal* (Anatomical Supplement) 34 (4): 64-9

TEACHING MATERIALS

1924 Peking Union Medical College Department of Anatomy. I. Description of the building and equipment. II. Instruction and research. *Methods and*

Problems of Medical Education, First Series, pp. 25-39. New York: The Rockefeller Foundation
1926 Outline of Laboratory Work in Neuro-anatomy. 2nd ed. Peking Union Medical College Press (1st ed. privately printed in 1923)

PREHISTORIC PHYSICAL ANTHROPOLOGY

1925 Human skeletal remains from Sha-Kuo-T'un cave deposits in comparison with those from Yang Shao Tsun, and with recent North China skeletal material. *Palaeontologia Sinica*, Series D, 1 (3)
—— A note on the physical characters of the prehistoric Kansu race. *Memoirs of the Geological Survey of China*, Series A, 5: 52-6
—— The aeneolithic Yang Shao people of North China. A brief resumé of the work done and in progress on the physical characters of this ancient people, their distribution and their apparent ethnic relationships. *Transactions of the 6th Congress of the Far Eastern Association of Tropical Medicine*, Tokyo, pp. 1111-14 (abstract)
—— Recent work in the field of prehistoric anthropology in China. Proceedings of the Joint Conference of the China Medical Association and the China Branch, British Medical Association, section on Anatomy and Anthropology, January 1925, Hongkong. 1 p. in *Black: Collected Papers. Volume II.* Library of the American Museum of Natural History, Osborn Library of Vertebrate Palaeontology
1928 A study of Kansu and Honan aeneolithic skulls and specimens from later Kansu prehistoric sites in comparison with North China and other recent crania. Part I. On measurement and identification. *Palaeontologia Sinica*, Series D, 6 (1)

PALEONTOLOGY

1922 The progress of the Third Asiatic Expedition of the American Museum of Natural History during the early part of this season's work. *China Medical Journal* 36 (4): 342-3 (abstract). Proceedings of the Anatomical and Anthropological Association of China
1925 Asia and the dispersal of primates. A study in ancient geography of Asia and its bearing on the ancestry of man. *Bulletin of the Geological Society of China* 4 (2): 133-83
1926 Tertiary man in Asia: the Chou Kou Tien discovery. *Nature* 118: 733-4 (also in *Science* 64: 586-7)
1927 On a lower molar hominid tooth from the Chou Kou Tien deposit. *Palaeontologia Sinica*, Series D, 7 (1)

18 Homo erectus

—— Further hominid remains of Lower Quaternary age from the Chou Kou Tien deposit. *Nature* 120: 954 (also in *Science* 67: 135-6)

—— (with E. Licent and Teilhard de Chardin). On a presumably Pleistocene human tooth from the Sjara-osso-gol (southeastern Ordos) deposits. *Bulletin of the Geological Society of China* 5 (3-4): 285-90

1929 Preliminary note on additional *Sinanthropus* material discovered in Chou Kou Tien during 1928. *Bulletin of the Geological Society of China* 8 (1): 15-32

—— Preliminary notice of the discovery of an adult *Sinanthropus* skull at Chou Kou Tien. *Bulletin of the Geological Society of China* 8 (3): 207-30

—— *Sinanthropus pekinensis*: a further note on the new material recovered at Chou Kou Tien in 1928 and its zoogeographical significance. *Proceedings of the Fourth Pacific Science Congress*, Java, 3: 105-12

—— *Sinanthropus pekinensis*: the recovery of further fossil remains of this early hominid from the Chou Kou Tien deposit. *Science* 69: 674-6

1930 Interim report on the skull of *Sinanthropus*. *Bulletin of the Geological Society of China* 9 (1): 7-22

—— A preliminary report on the discovery of a skull of adult *Sinanthropus pekinensis* at Chou Kou Tien. *The China Journal* 12: 163-4

—— Discovery of skull of adult Sinanthropus at Chou Kou Tien. *Anthropologischer Anzeiger* 6: 358-9

—— Notice of the recovery of a second adult *Sinanthropus* skull specimen. *Bulletin of the Geological Society of China* 9 (2): 97-100

1931 Palaeogeography and polar shift; a study of hypothetical projections. *Bulletin of the Geological Society of China* 10 (Grabau Anniversary Volume): 105-57

—— On an adolescent skull of *Sinanthropus pekinensis* in comparison with an adult skull of the same species and with other hominid skulls, recent and fossil. *Palaeontologia Sinica*, Series D, 7 (2)

—— Evidences of the use of fire by *Sinanthropus*. *Bulletin of the Geological Society of China* 11 (2): 107-8

—— Preliminary report on the *Sinanthropus* lower jaw specimens recovered from the Chou Kou Tien cave deposit in 1930 and 1931. *Bulletin of the Geological Society of China* 11 (3): 241-6

1932 Skeletal remains of *Sinanthropus* other than skull parts. *Bulletin of the Geological Society of China* 11 (4): 365-74

1933 (with Teilhard de Chardin, C.C. Young, and W.C. Pei; edited by Davidson Black). Fossil man in China. The Choukoutien cave deposits with a synopsis of our present knowledge of the Late Cenozoic in China. *Memoirs of the Geological Survey of China*, Series A, 11

―― The brain cast of *Sinanthropus* – a review. *Journal of Comparative Neurology* 57 (2): 361-8
―― On the endocranial cast of the adolescent *Sinanthropus* skull. *Proceedings of the Royal Society*, Series B, 112: 263-76
1934 On the discovery, morphology, and environment of *Sinanthropus pekinensis*. *Philosophical Transactions of the Royal Society of London*, Series B, 223: 57-120 (Croonian lecture 1932)
―― Present state of knowledge concerning the morphology of *Sinanthropus*. *Proceedings of the Fifth Pacific Science Congress*, Victoria and Vancouver, B2, 1: 2711-13. Toronto

2

HARRY L. SHAPIRO

Davidson Black: an appreciation

Le nom de Davidson Black est aussi inséparable des fossiles de l'homme de Pékin que le sont ceux de Louis Leakey, Raymond Dart et Robert Broom des fossiles découverts en Afrique ou que celui d'Eugène Dubois des fossiles pithécanthropes de Java. Davidson Black n'a pas découvert les fossiles de l'homme de Pékin, mais c'est lui qui a su en reconnaître et décrire la véritable importance et rassembler les preuves qui ont confirmé l'hypothèse. Lorsqu'il est venu enseigner à Pékin en 1919, il était la personne toute désignée pour faire une évaluation bien documentée de ces trouvailles passionnantes. Il avait reçu, à Toronto, une formation en médecine, en anatomie, et en biologie et il avait par la suite étudié la neuroanatomie et l'odontologie comparatives avec des spécialistes. La géologie structurale et stratigraphique présentait pour lui un intérêt fondamental que l'on trouve déjà dans les expériences qu'il fit pendant son adolescence au Canada. Et jusqu'à sa mort, survenue à l'âge de 49 ans, il éprouva pour l'anthropologie physique une passion que très peu de contemporains partageaient à ce point. Davidson Black était fort estimé, tant sur le plan professionnel que personnel, de ses collègues du collège médical de l'union à Pékin et des scientifiques des quatre coins du monde qui, à son époque, s'intéressaient à l'étude de l'évolution de l'homme.

In the history of the study of human evolution there are a series of associations that have become fixed in the minds of all of us who know something of the subject. *Pithecanthropus erectus*, for example, and his discoverer, Eugene Dubois, are inevitably linked. Raymond Dart and *Australopithecus africanus* go together, as does Robert Broom with *Paranthropus* and *Plesianthropus*, and Louis Leakey with *Zinjanthropus* and *Homo habilis.* There is still another association that has world-wide acceptance. This is Davidson Black and Peking Man.

In many, but not in all of these associations, the linkage is partly the result of the acclaim that comes to the discoverer of an important fossil. In a few instances, however, it is the consequence of the recognition and identification of a fossil's true significance and the marshalling of the evidence to demonstrate this judgment. Davidson Black belongs in this category. He was not responsible for the discovery of Peking Man. That was the result of the efforts of J. Gunnar Andersson, who planned the explorations at Chou Kou Tien, and of Otto Zdansky, Birger Bohlin, and Pei Wen-Chung among others, who did the field work. But when Zdansky at Uppsala, where the Chou Kou Tien fossils were sent for study and identification, found among the miscellaneous fragments a couple of teeth that seemed to have primate or possibly hominid features, they were returned to Peking and referred to Davidson Black for his judgment. This was in 1925-26.

Why was Black chosen for this demanding task? He had come to Peking in 1919, shortly after the end of World War I, to join the Department of Anatomy at the newly founded Peking Union Medical College. Black had, to a considerable extent, chosen to accept this post in Peking because of his interest in human evolution. He had a conviction, influenced by the hypothesis of William D. Matthew, a fellow Canadian, and Henry Fairfield Osborn, that man had originated in Asia (Matthew 1915). Thus, being located in Peking would have given him the opportunity to explore likely areas for traces of early man. His tentative searches, however, had until 1925-26 not yielded any concrete evidence of early man. In the seven years since his arrival Black had already so firmly established his knowledge and authority in the fields of dental evolution and the comparative osteology of man, that he was the natural and obvious choice for a reliable evaluation and assessment of these exciting new finds. In this unexpected way, his choice of Peking as the seat for his research had paid off.

With only a few exceptions, the wisdom in turning to Black became immediately apparent to his colleagues. But in scientific centers around the world, where these matters were of prime importance, this recognition was not so prompt. Black published in 1926, in *Nature* and in *Science*, his conclusion that these teeth were hominid. Subsequently, he went even further to identify them as relics of a new type of early man, the oldest ever found in China or the Asiatic mainland, going back in age to the beginning of the Pleistocene, possibly, as dated then, to a million or more years ago. Considering the paucity of the evidence, it was with notable confidence that he ventured in 1927 to give it the distinctive name of *Sinanthropus pekinensis*.

I had just completed my graduate studies at the time and remember distinctly the wave of incredulity with which this official announcement was received by various authorities around the world and in particular the skeptical discussions I had with my colleagues and teachers at Cambridge.

But when in 1928 cranial and mandibular fragments and in 1929 a well preserved skull completely vindicated Black's identification, an equally exciting wave of surprise circled around. This time, however, the excitement was permeated with admiration for the perception and skill of Davidson Black.

It is of some interest, and this fits a pattern that I shall deal with presently, that Black's outstanding knowledge of dental anatomy that enabled him to evaluate with precision the significance of the first dental fossils was not a matter of chance. He had, on coming to Peking, begun to make an outstanding collection of dental casts which he had studied with considerable intensity. This preparation for a field of research for which he had little encouragement from available material reflected a commitment to the problem of human evolution that went back to an early stage of his career.

Let me briefly outline his biography. Davidson Black was born on July 25, 1884, in Toronto, the second son of Davidson Black Sr, QC, and Margaret Bowes Delamere. Two years after his birth, his father died of a heart attack at the age of 48, strangely prophetic of his son's death in 1934 at almost the same age and from the same cause. Paul H. Stevenson (1934) wrote of the 'bravery and spirit' with which his mother — faced with the problem of providing for her two sons — accomplished her task. She was of Irish origin from a family she proudly traced back to Rollo, the Dane, and Fitz-Hubert De-la-Mer, named for his maritime distinction. Again, according to Stevenson, our Davidson Black treasured this long genealogy of his family.

As a boy Davidson Black spent his summers with his maternal uncles pursuing his love of exploration and adventure on the Kawartha Lakes of Canada where he became very well versed in natural history and nature lore. It was here that he acquired the skill of an outstanding canoeist. This enabled him later during high school summer vacations to take a job hauling supplies by canoe for the Hudson Bay Company. These solitary trips covering thousands of miles through wild, uninhabited country proved a notable test of the self-reliance and skill of the youthful Black. On one occasion he faced the horror of an encircling forest fire by standing for a day and two nights in a lake immersed up to his neck.

Stevenson attributed Black's fundamental interest in structural and stratigraphic geology to this first-hand experience with the landscapes he explored in those formative years of his life; he assures us that Black's geological and paleontological colleagues were deeply impressed by his knowledge in these fields.

In 1906 Davidson Black received his medical degree from the University of Toronto and then, because of his profound interest in biology, he returned to that university to study the subject, receiving a BA in 1909. After completing his academic training Black taught anatomy at Western Reserve University in Cleveland.

But the characteristically explorative temperament of Davidson Black had developed still another field of interest — neuroanatomy. In 1914 he went abroad to study with Professor Ariëns Kappers, the distinguished Dutch authority in this subject, and with Sir Grafton Elliot Smith in England. Black had already published several papers on neuroanatomy and his experience with these two leading authorities reinforced this interest. But apparently it was Elliot Smith who also turned the direction of this interest toward the neuroanatomy of prehistoric man. Smith was at the time deeply involved with the recently discovered Piltdown Man, having engaged in a rather heated controversy on its reconstruction and morphological significance (Smith 1913; Keith 1913, 1920). It was perhaps inevitable, in retrospect and considering Davidson Black's latent interest in brain morphology, that he should have become intellectually captivated by the problems of human evolution this fossil presented. It is also ironic, but of no significance in this context, that the famed Piltdown fossil turned out to be a fake. In any event Black had a taste of the fascinating problem of deciphering the clues to be solved in the study of human evolution.

But this beginning was interrupted by World War I in which Black served. It was not until the war was over and he had accepted (in 1919) the post of professor of neurology and embryology at the newly founded Peking Union Medical College that he was once again ready and eager to resume his interests in this field.

This new post came to him through his college friendship with Edmund Vincent Cowdry, who was appointed professor of anatomy in 1918, the first professional assignment in the college. Dr John Z. Bowers (1972) tells the story that when Black arrived in Peking in 1919 he and his family were met by the Cowdrys who escorted them to his residence in the Peking Union compound. When they arrived at the house, they found to their dismay that there were no doors, no window frames, and no furniture. The Cowdrys then hustled the Blacks off on a sightseeing trip around Peking but apparently also left some orders behind. On their return it was reported that the miraculous efficiency of the Chinese had in the short interval corrected all the deficiencies, and the Blacks found doors, windows, and furniture all appropriately installed and a servant smiling as he greeted them.

Bowers also reported that one of Black's first research projects was in collaboration with Bernard Emms Read. In a letter that Black wrote December 5, 1919, and cited by Dora Hood (1964:49), he described the project as follows: 'During the last month I have had a large female camel living in the basement. Read, our physiologist chemist, is doing the nitrogen metabolism on this beast and I shall take what remains.'

Black, fired by his deep interest in anthropology and human evolution, almost immediately began to assemble comparative material that would enable him to pursue this interest. In these first years, convinced as he was that the early hominids had emerged from their primate ancestors in Asia, he actually made several exploratory expeditions and planned others for the future.

This concern and commitment to the study of human evolution, with all the research it entailed in comparative osteology and odontology, also opened up some related research areas in physical anthropology that can be seen as a spin-off from his central interests. In 1920 he published a paper on the anthropometry and observations of healthy subjects. This was followed by a series of craniological studies of the Sha Kuo T'un and the Yang Sha Tsun series of prehistoric Chinese. And in 1925 he published an analysis of the so-called prehistoric Kansu race.

These researches, although rarely mentioned, were highly valuable for Black in assessing the morphology of the Peking fossils and also in establishing his competence and reputation as the obvious choice to deal with the Peking fossils when they were discovered. In any case, there was no one else in Peking as qualified to do the job.

It is of passing interest that Black's activity in physical anthropology in the Department of Anatomy raised some questions in the administration, which was somewhat disturbed that he was pursuing little or no conventional anatomical or neurological research. There even was talk of replacing him. In fact, however, he was appointed professor of anatomy, abandoning his neurology assignment, and no further reservations were expressed about the field of research he had elected to follow.

Thus we see that in 1926, when Peking Man appeared ready for authoritative recognition, Black was there, prepared and eager for the task that he performed with the insight he had developed from his exploratory investigations. In a sense, chance and luck had coordinated Black's deepest interests and commitment with a discovery in which he had no part.

Black's achievement in connection with Peking Man led to a world-wide recognition. In 1932 he was elected to a fellowship in the Royal Society, and shortly after his death he was posthumously awarded the 1931 Elliot Medal by the National Academy of Sciences with the following citation by Henry Fairfield Osborn: 'Award given for the brilliant series of papers on Peking Man, culminating in the paper "On an adolescent skull of *Sinanthropus pekinensis* in comparison with an adult skull of the same species and with other hominid skulls, recent and fossil." ' He also received other awards and recognition from scientific societies around the world and was easily one of the outstanding figures in the field of human paleontology.

I cannot speak of Black as a person from any long or deep association with him. I did meet him in the winter of 1931–32 when I was visiting Peking and I can confirm from this brief contact the uniform reaction of his old friends and colleagues at Peking that he was an extraordinarily kind and warm individual. He received me very cordially, inviting me to his laboratory and demonstrating with great care and detail his techniques of preparing the fossil specimens for study. I remember the clouds of dust that surrounded his head and face as he showed me how he used a dental drill to remove incrustation from the fossil fragments. When I asked him if he wore a mask, he laughed and dismissed the idea.

Black was an extremely dedicated worker. His routine was to work all night so that he would not be interrupted and to sleep during the day. Even toward the end, when he knew that his health was failing, he continued this regime with the intense devotion he had given to his researches in the past. Dr V.K. Ting (1934) reported that he did not even let his wife know of his serious condition, maintaining to the end a cheerful and optimistic demeanor.

One aspect of Davidson Black's character that I find profoundly revealing has rarely been mentioned. I shall quote Dr Ting (1934: 18-19) who, as honorary director of Cenozoic Research in China, knew him well:

The last point which I should like to touch is a delicate one, but I am going to touch it nevertheless. It is frankly admitted that sometimes we find cooperation between Chinese and foreigners in scientific work rather difficult. The reasons I think are not difficult to seek. Firstly, many foreigners are suffering from a superiority complex. Subconsciously they think somewhat like this; here is a Chinese, he knows something about science, but he is a Chinese nevertheless — he is different from an European, therefore we cannot treat him in the same way. At best his manners become patronizing. On the other hand, their Chinese colleagues are suffering from an inferiority complex. They become self-conscious and supersensitive, always imagining that the foreigner is laughing at them or despising them. Ninety per cent of the trouble between Chinese and foreign colleagues working together comes from these two factors. In my dealings with Davidson Black, and I think Black's colleagues will bear me out, I never found him suffering from such a complex, and his Chinese colleagues became also free from theirs. In politics Black was a conservative, but in his dealings with his Chinese colleagues, he forgot altogether about their nationality or race, because he realized that science was above such artificial and accidental things. This I think is an example for all of us to follow. Davidson Black is dead. Long live his memory! In the life and work of those who have come into contact with him, he has achieved immortality!

3

G.H.R. VON KOENIGSWALD

Davidson Black, Peking Man, and the Chinese Dragon

Les 'dents de dragon,' dents fossiles de mammifères d'origine chinoise, se vendent comme amulettes dans les pharmacies en Chine comme ailleurs. Dans une vaste collection assemblée par Haberer dans le nord de la Chine en 1900, se trouvait une dent humaine recueillie à Pékin. Cette dent concentra l'attention sur les environs de Pékin et mena à la découverte du célèbre site de l'homme fossile à Chou-Kou-tien. Sous la conduite éclairée de Davidson Black, Birger Gohlin et W.C. Pei recueillirent les restes de l'homme de Pékin.

À la suite des nombreuses enquêtes menées dans les pharmacies chinoises, principalement dans le sud de la Chine, il a été démontré que la plus grande partie du matériel fossile provient de deux niveaux stratigraphiques distincts. Le premier groupe est fait de dents mieux fossilisées, qu'on appelle donc 'dents de dragon de première qualité.' Ce groupe est d'âge pliocène et contient d'abondants vestiges d'hipparions, de girafes, d'antilopes, d'hyènes, mais aucun reste de primate. L'autre groupe, les 'dents de dragon de seconde qualité' remonte au pléistocène. L'hipparion en est absent, mais l'assemblage renferme des restes de cerfs, d'ours et de tapirs en quantité. Les primates sont représentés par des vestiges de singes, de gibbons, d'orang-outans, de *Gigantopithecus blacki* et par des dents isolées appartenant à l'homme de Pékin. Toute la faune provient du sud de la Chine.

With the discovery of Peking Man, Davidson Black became world famous. For first time, fossil skulls and jaws of a genuine primitive 'human' being had been found in China. However, these discoveries in the countryside of Peking were not completely unexpected. Around 1900, some 25 years before Davidson Black's identification of *Sinanthropus pekinensis*, a human tooth had already been obtained in a Peking drugstore. Although no one could be sure that this mysterious tooth came from Peking, the find serves to illustrate an important 'source' of fossil remains in China.

To the Chinese people, the dragon is real. It is real because they find 'dragon' bones and teeth in many places of their country. These fossils are frequently the remains of large animals, such as horses, rhinoceroses, and elephants, and the people have felt that the remains of such mighty animals must have great medical power. Thus, the peasants have collected these remains and have sold them to drugstores. Unfortunately for the paleontologist, the dragon teeth, 'lung tse,' are regarded as having greater medicinal value than the dragon bones, 'lung ku.' Therefore, skulls and jaws are generally broken to separate the teeth. Because the material is very expensive, sold according to weight, and used only in small quantities, whole teeth are often broken as well. The breaking serves another purpose for it results in exposing the pulp cavity which often shows small crystals of calcite due to fossilization, proof that the item is a genuine dragon tooth of first quality.

Many of the 'dragon' remains are of Pliocene fauna and occur in veritable 'bone beds,' the long teeth of *Hipparion* sometimes serving as their trademark. *Hipparion* is a three-toed horse and the index fossil of the Pliocene period. Its remains are common; over the years I have seen between 30,000 and 50,000 isolated teeth. They can be bought in practically every ordinary Chinese drugstore, even those in small Chinese communities outside the country. Under such circumstances, we have found dragon teeth all over Southeast Asia, from Bangkok to Singapore, across Indonesia, and up to the Philippines. We have seen the teeth in San Francisco, New York, and Honolulu, and even in Canada. There are several such drugstores in Toronto, not far from City Hall, and we have found one on Pender Street in Vancouver in the back of a modern shopping center.

Fossils from Chinese drugstores have been known for a long time. Dr K.A. Haberer, a German naturalist who had travelled in northern China between 1899 and 1901, was, however, the first to make a systematic collection of dragon teeth. He did so not to amass 'souvenirs,' as has been stated in one of the books about Davidson Black (Hood 1964), but on behalf of Professor von Zittel in Munich, his teacher and the first paleontologist to write a treatise of paleontology. Haberer's collection was described by Max Schlosser (1903) in a monumental publication containing 14 large plates. The bulk of the collection came from Pliocene deposits but also contained material from the Pleistocene epoch. The rich Pliocene fauna contained the teeth of bears, different hyaenas (one as large as a calf), otters, saber-toothed tigers, beavers, mastodons, different rhinoceroses (including a new species, *Rhinoceros habereri* Schlosser), *Hipparion* (more than 600 teeth), a large *Anchitherium* (another three-toed horse but with low-crowned teeth), pigs, a very large camel, giraffes, different species of deer, and antelopes.

From Haberer's collection many species were described for the first time and gave us an idea about the richness of the fossil Chinese fauna. But the prize

FIGURE 3.1
'Haberer's tooth,' an upper human molar from a drugstore in Peking. Actual size. (After Schlosser 1903.)

specimen of this collection was the aforementioned human tooth from a drugstore in Peking (Figure 3.1). It is a small upper molar, very worn and with fused roots. Some red earth is still attached to the roots. Because this condition appears with *Hipparion* teeth, as well as with other Pliocene fossils, and the preservation is generally similar, Schlosser suggested that the human tooth might be of Pliocene age. But Schlosser was cautious; he referred to the tooth as '?Anthropoide g. et sp. indet?' However, the last sentence of his long discussion reads as follows: 'Der Zweck dieser Mitteilung ist, spätere Forscher, denen es vielleicht vergönnt ist, in China vielleicht Ausgrabungen vorzunehmen, darauf Aufmerksam zu machen, dass dort entweder ein neuer fossiler Anthropoide oder der Tertiärmensch oder doch ein Altpleistozäner Mensch zu finden sein durfte' (Schlosser 1903:21). Thus Schlosser, already in 1903, expressed the hope that future scientists would find in China a new anthropoid, either a man of Tertiary age or a hominid from the Lower Pleistocene. This was almost a prophecy, for the hope was fulfilled 25 years later by Davidson Black.

Black must have been very much intrigued by the 'dragon bones.' He published a map of their locations according to the information of the *Pen T'sao*, the old Chinese pharmacopoeia. The last official edition of these writings is from 1597, but some of the records on which they are based go back to the Wei Period, 7th century BC. I am here reproducing this unique map according to Read (1934) who has made a special study of Chinese drugs (Figure 3.2).

China and its antiquities generated much interest in Sweden following that country's dispatch of missionaries to China. As a result, the Swedish Academy sent a number of scientists to China in the 1920s to collect fossils. This began the fine collections now at the University of Uppsala. Shortly thereafter the Swedish match king, Ivar Kreuger, founded the *Palaeontologia Sinica* for the publication of paleontological and paleoanthropological papers resulting from Chinese studies. It was in this series, of course, that Black's and Franz Weidenreich's famous studies on Peking Man were published. At its inception Black's name appeared as one of the 'Honorary Research Associates' (Figure 3.3).

30 *Homo erectus*

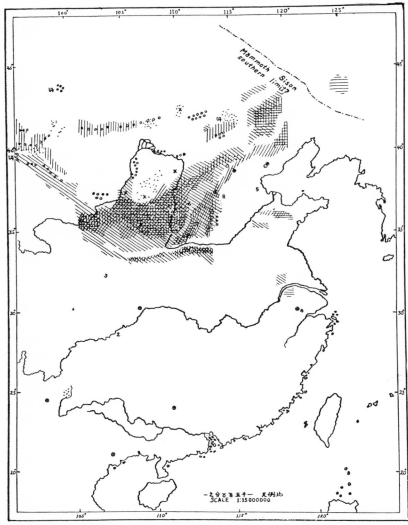

1. 益州(四川成都府廣漢縣) Yi Chou (Kuang Han Hsien, near Chengtu, Szechuan)   2. 巴州(四川巴縣) Pa Chou (Pa Hsien, Szechuan)
3. 梁州(陝西南鄭縣) Liang Chou (Nan Cheng Hsien, Shensi)   4. 劍州(浙江天台縣附胞) Yen Chou (Near Tien Tai Hsien, Chekiang)
5. 滄州(河北滄縣) Tsang Chou (Tsang Hsien, Hopei)   6. 太原(山西陽曲) Tai Yuan (Yangchü, Shansi)   7. 龍門(山西河津陝西韓城之間)
Lung Men (Between Hochin, Shansi, & Hancheng, Shensi)   8. 鎮州 (河北正定諸縣) Chen Chou (Chengting, Hopei.)

Distribution of DRAGON BONES in China,
Fossil fauna from the formations of,
Lower Pliocene, Late Pliocene.
⊕ Early Pleistocene, Late Pleistocene.
Compounded from "Fossil Man in China", by Davidson Black,
The numbered places refer to the Pen T'sao records.

THE GEOLOGICAL SURVEY OF CHINA
PALÆONTOLOGIA SINICA
中 國 古 生 物 誌

Palæontologists to the Geological Survey: Chief Palæontologist: A. W. Grabau:
Palæontologists: Y. C. Sun, C. Ping, T. C. Chow, Y. T. Chao (deceased).
C. C. Tien, C. C. Young, C. C. Yü, S. S. Yoh, K. H. Hsü, W. C Pei, T. K. Huang, Y. S. Chi.

EDITORS:
V. K. TING AND Y. C. SUN

FOUNDER
提 倡 人
Ivar Kreuger, Sweden
克魯格　瑞典

PATRON
贊 助 人
Sinyuan Daw King, Chekiang
金繇沁園浙江吳興前清中書科中書

HONORARY RESEARCH ASSOCIATES
名 譽 研 究 員
Carl Wiman, Upsala
維曼　瑞典
T. G. Halle, Stockholm
赫勒　瑞典
J. S. Lee, Shanghai
李四光　上海
Davidson Black, Peiping.
步達生　北平

FIGURE 3.3
Davidson Black as 'Honorary Research Associate' of the *Palaeontologia Sinica*.

In connection with the Swedish research program, Otto Zdansky was first to dig in Chou Kou Tien in 1921 and 1923. He also found the first human teeth: a very worn upper molar, already recognized in the field, and a first premolar, recognized while the collection was in Sweden. Zdansky cautiously described the

---

FIGURE 3.2
Davidson Black's distribution of fossil mammals according to the information of the old Chinese pharmacopoeia. (After Read 1934.)

32 Homo erectus

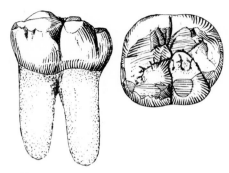

FIGURE 3.4
The first tooth of Peking Man, a lower first molar found by Birger Bohlin. It is very long and broad, and also taurodont. Twice natural size.

two teeth as '?*Homo* sp.' (Zdansky 1928:131). He found a second lower premolar in his collections in 1952 (Zdansky 1952). That human remains might be expected at Chou Kou Tien was first predicted by J. Gunnar Andersson. He suspected that strange sharp fragments of quartz, not infrequent at the site, might be primitive implements of early man.

After Zdansky had left, Davidson Black, with the help of the Rockefeller Foundation, continued work at Chou Kou Tien. His collaborator was Dr Birger Bohlin, a young and enthusiastic Swedish paleontologist. On October 16, 1927, a large human molar (see Figure 3.4) was found by Bohlin, who hastened to bring his precious find directly to Peking to personally hand it over to Black. Black directly recognized that this remarkable tooth was unlike any other known human molar and described the find as *Sinanthropus pekinensis* (Black 1927). Peking Man was born! When later the skulls came to light it became evident that Black was right, as some scientists had doubted that just one tooth was enough to recognize a new type of early man. With the skulls he could demonstrate that Peking Man was closely related to the famous Java ape-man, *Pithecanthropus*. Eugene Dubois, a Dutch physician, had discovered in 1891 near Trinil in Central Java a flat skull cap with a heavy torus above the eyes. For many scientists the find was not enough to prove the human nature of this much disputed fossil. By Black's discovery the discussion had come to an end. *Pithecanthropus*, too, was undoubtedly a primitive hominid.

Enthusiastically, I published my first paper on Black's discovery (1931). But most disappointing for us, Dubois would never admit any relationship between Java Man and Peking Man, and even went so far as to suddenly declare that 'his' *Pithecanthropus* was not human at all, but merely a giant gibbon as the find had been referred to by some of his adversaries. He had lost his ape-man monopoly after nearly 40 years, and this apparently was too much for him. A new skull, *Pithecanthropus* II, which we found in 1937 at Sangiran, Central Java, and more complete than his own find, he declared to be a fake.

Now, after World War II, all the precious remains of Peking Man have been lost. It is not necessary to repeat the sad story here. We also lost part of our fossil collections during the Japanese occupation in Java. Some boxes with fossils simply vanished. Nobody could really be interested in fossil teeth and bones, but they are gone nonetheless. It is the same with the Peking fossils, and I do not believe all the romantic stories about mysterious ladies and boxes filled with human bones (see Janus 1975). But one remark is in order. With the help of my wife and some Swedish and Swiss friends we were eventually able to save all our human fossil material from Java. After the war, at the American Museum of Natural History in New York, we divided our finds for study. Professor Franz Weidenreich took charge of the Solo skulls, also discovered in Java, while I studied my *Pithecanthropus* remains. One day we were visited by a Professor Watson of London. In Weidenreich's office Watson spotted the Solo skulls, and being a paleontologist he recognized them as original specimens. He asked no questions but mistook them for the skulls of Peking Man. Back in London he told his students that he had seen Weidenreich and the Peking skulls in New York. When this bit of information got into the newspapers, we had the greatest difficulty trying to tell the journalists that this had been an error. Nevertheless, the story about Peking Man in New York came into being and can still be found in modern publications in China.

It was in 1937 that I came to Peking for the first time. C.C. Young, an old friend since my student days in Munich, took me to Chou Kou Tien. The roads were bad, but there was a small and uncomfortable local train. The site was a large quarry, long and quite narrow, about 20 m below surface. The walls were neatly painted with white lines, subdividing the place into quadrangles on account of the excavation. On a hill nearby was a small temple, which had been moved there from its original site at the place of the excavation. The inner walls of the temple had been newly painted by a local artist who had given a rendition of our activities.

During my visit I naturally learned a lot about Davidson Black. He had made many friends, had lived a very social life, and had spent many evenings in the Peking Hotel. But around midnight he used to disappear into his office, working until the early hours of the morning, then going home only to reappear around noon. In 1939 we spent five months in Peking, working together with Weidenreich on a comparative study of *Pithecanthropus* and *Sinanthropus*. But due to the war our results were not published, except for a short article in *Nature* (von Koenigswald and Weidenreich 1939). On March 15, the anniversary date of Davidson Black's death, the whole Department of Anatomy of the Rockefeller Medical Center – the Jüh Wang Fu – with Professor Weidenreich and all his assistants went out to the little clean European cemetery of Peking to decorate Black's grave with fresh flowers to commemorate the man who had contributed so much to the science of early man.

Because of the later Japanese occupation it was not possible to go out to Chou Kou Tien again, but we still kept contact with the people there. In 1973 we went to Peking on a most generous invitation of the Academia Sinica. We found Chou Kou Tien greatly changed. The road to the place had been paved and, being in good condition, we went by car. The site had been made a national monument with signs and explanations all over the place. There was an excellent little museum, exhibiting casts of the original skulls, and very good reconstructions by Wu Ju Kang of Peking Man himself and, on a small scale, of his life and of the animals that had lived at the time. Many original fossils were on exhibit, from Chou Kou Tien and from other places, and there was even the fine cast of a large dinosaur. There were many visitors, including apparently whole schools of children who came here by bus to have a glimpse of China's earliest history.

My own search for dragon teeth began early in 1931 when I came to Java to serve as a paleontologist to the Geological Survey. I had finished my studies in Munich, had known old Professor Schlosser quite well, and was familiar with the Haberer Collection. In Java the Chinese community was large, and the people lived exactly as in China. There were even typical Chinese drugstores along Pasar Baru in Bandung, where we lived, but it took me sometime to discover what I was after. First I had asked for 'gigi binatang,' teeth of animals, until a friendly old Chinese told me that these were 'Dragon Teeth.' Once I had a prescription, I found them everywhere, as I have already told in the beginning.

This had been the first time that somebody had looked for this mysterious medicine outside of China. I had hoped to discover by this most romantic method a Pliocene Man, but while there was quite a lot of Pliocene material on the market, I never found a single tooth of a fossil primate in this assemblage, not even a common monkey.

But besides the Pliocene fossils there was a different kind of material on the market (Figures 3.5 and 3.6). It came, as we learned, via Hong Kong and Canton where Haberer had not collected. The original sites, as we were told, were caves and fissure fillings in the southern Chinese provinces of Kwangsi and Kwantung. On Black's map this region is blank. In these provinces there are many limestone formations that show typically Karst weathering and dissection. The neighborhood of Kweiling is one of the most romantic landscapes in Asia and has inspired Chinese painters from the Han Period on.

The drugstore material from the south was different in age, preservation, and faunal content. The teeth were not as heavily fossilized as those from the Pliocene; it seems that they were regarded as a kind of 'Dragon Teeth — second quality' and were apparently younger. Teeth of elephants were proof of Pleistocene age. The preservation was most unusual in that all bony parts, including the roots, had been gnawed away by large porcupines who need the lime for their quills. They damaged as well our hope of ever getting complete jaws and skulls from these sites, or only as exceptional specimens. One site is known where the bones have been preserved, but this is in Szechuan.

Species for species, the composition of the fauna is different from the Pliocene assemblage. There is neither *Hipparion*, nor even modern horse. Many bears of modern type, dog and tiger, and teeth of the Giant Panda are not as rare, but the Lesser Panda is practically absent. There are rhinoceroses, pigs, and deer. Also present are the teeth of *Stegodon* (a primitive elephant with low-crowned molars) and those of true elephants, but there are no mastodon remains. However, and most important, there are many teeth of various primates including Man!

There are, to begin with, two species of common monkeys, a larger form (*Macaca* cf. *robusta*) and a smaller one, the latter probably identical with the 'Golden Monkey' of China. There are a number of isolated teeth of the gibbon, certainly representing two species. Then, to our surprise, we found quite a series of teeth of the orangutan (*Pongo*), still living in the south on the islands of Sumatra and Borneo. But most of the Chinese orangutan teeth are much bigger than those of the modern form, some of them even surpassing the large teeth of the gorilla, the most powerful of the living anthropoids. We collected more than 1,500 teeth, so this animal must have been quite common. Interestingly some of the teeth we found differed from the orangutan teeth in having less wrinkles and a double cusp on the middle of the upper molar. These teeth are so similar to certain ones of *Australopithecus* from Swartkrans in South Africa, that they most probably indicate an australopithecoid form in China, which we have called *Hemanthropus*. A few isolated teeth and no indication of the site is all we have at this moment. In Java we have similar problems with *Meganthropus*.

We also have a large selection of human teeth. Most of them are recent or subrecent. Some have, as x-ray pictures show, a very high pulp chamber. They are as taurodont as some of the famous Neanderthal teeth from Europe, especially those from the classical site of Krapina. Whether they indicate a kind of Neanderthal Man from China, we do not know. Other teeth are very small, like the first tooth obtained by Haberer in a drugstore in Peking. The same type of teeth are also to be found in a small fragment of a jaw with some red clay attached. Schlosser, just by means of the red clay on the Peking drugstore tooth,

36 *Homo erectus*

FIGURE 3.5
Selection of 'Dragon Teeth' from drugstores in Hong Kong and Canton. About two-thirds actual size. (Coll. v. Koenigswald; photo: A. Hoffman/Senckenberg.) Key: see Figure 3.6.

was inclined to discuss its Pliocene age. Father Teilhard de Chardin, who studied my collection, assured me that these teeth must come from Mesolithic layers, often present in Chinese caves. Hence, Haberer's tooth from Peking must belong to a Mesolithic *Homo sapiens*, and it was after all a modern tooth which touched off the hunt for Peking Man.

But there are other human teeth which are well fossilized. Among them are two first lower premolars of considerable size, more than twice as large as the

37 G.H.R. von Koenigswald

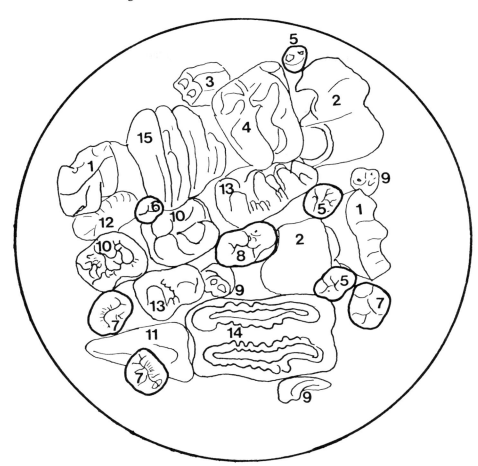

FIGURE 3.6 Key to Figure 3.5
A Pleistocene assemblage: 1, deer; 2, rhinoceros; 3, mountain goat; 4, giant tapir; 5 *Sinanthropus 'officianalis'*; 6, *Homo sapiens* ('Haberer' type); 7, fossil giant orang; 8, *Gigantopithecus blacki* v.K.; 9, monkey (*Macaca robusta*); 10, giant panda; 11, tiger; 12, bear; 13, pig; 14, Stegodon; 15, elephant.

corresponding teeth of modern man. Exactly such large premolars are typical for Peking Man from Chou Kou Tien. Thus in our collection, we have traces of Peking Man from southern China. There are in the molars some smaller differences from the classic form — no cingulum in the lower molars, Carabelli's pit in the upper — so we decided to preliminarily assign a new name,

38 Homo erectus

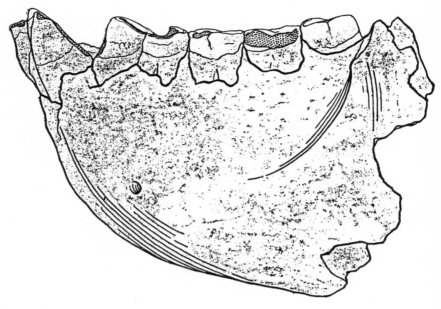

FIGURE 3.7
Side view of the *Gigantopithecus blacki* mandible III. This is the largest mandible of a higher anthropoid ever found; from isolated teeth it must be deduced that there are still larger jaws. About three-quarters actual size. (After Woo 1962.)

Sinanthropus officinalis, '*Sinanthropus* from the drugstores' (von Koenigswald 1952).

The greatest surprise in the collection is a last lower molar, obtained in 1935 in Hong Kong. This tooth, larger than a gorilla's and the larger orangutan teeth, is markedly different from the orangutan teeth of the same drugstore. The tooth is higher, there are no wrinkles, and the cusps must have been depressed but are rather high and swollen. This unusual type must have belonged to an as-yet-unknown higher primate. I was so convinced that this must be the very first indication of a completely new species that on the basis of this single tooth I created a new genus and species. I have called the find *Gigantopithecus blacki* v.K., in honor of the discoverer of Peking Man (von Koenigswald 1935).

Weidenreich and others would not believe me and first referred the tooth in question to a giant orangutan. Later, Weidenreich changed his mind and even went so far as regarding *Gigantopithecus* as ancestral to *Homo sapiens*! The reduction of the human dentition in the course of evolution is evident, so why couldn't there have been a giant at the beginning? But the teeth (later we found more) are too overspecialized to fit into the *Homo* line. It was not before 1956, more than 20 years after we had spotted that creature in the drugstores, that Dr Woo, a member of the Chinese Geological Survey, found during excavations of caves in the Tahsin and Liucheng Districts in Kwangsi many isolated teeth and three enormous mandibles, firmly establishing and recognizing the existence of *Gigantopithecus blacki* (Woo 1962).

Gigantopithecus (Figure 3.7) was the largest higher primate that ever lived. The geological age is Lower to Middle Pleistocene. The skull is still unknown; according to our estimates (based on the size of the dentition) the brain capacity might have been as high as 700 cm^3 and, therefore, in a prehuman range. The position of this creature is still under dispute. For some it is an exceptional anthropoid (also my first impression, therefore the name *pithecus*), for others he might be an offshoot of the human line, an offshoot that acquired gigantic size. There is a Tertiary forerunner from the Indian Siwalik Hills, of Middle Pliocene age, and very recently a probably ancestral form of Lower Pliocene age has turned up in Europe (not yet described).

China is full of surprises, and so are the Chinese drugstores. In our experiences answers come straight from the Dragon's mouth.

ALAN MANN

The significance of the Sinanthropus casts, and some paleodemographic notes

Bien que les chercheurs n'aient pas accès aux fossiles originaux d'*Homo erectus* découverts dans la localité 1 de Chou-kou-tien, plusieurs musées, notamment l'American Museum of Natural History à New York, possèdent des collections quasi complètes de moulages en plâtre d'un modelé exquis, y compris des moulages de dents isolées. Ces répliques ne peuvent se substituer aux fossiles eux-mêmes, mais elles constituent d'importants spécimens d'étude qui devraient être mis davantage à contribution.

Les descriptions publiées au sujet de l'échantillon original de la localité 1 portent à croire qu'il existait un dimorphisme sexuel considérable chez ce groupe. Toutefois, l'étude des collections de moulages et des descriptions des fossiles originaux faites par Franz Weidenreich et Davidson Black révèle que le degré d'anomalie sexuelle a peut-être été exagéré. L'échantillon semble présenter une variabilité morphologique beaucoup plus grande qu'on ne l'avait laissé entendre jusque-là. Par ailleurs, le nombre de sujets représentés provenant de cette localité 1 se situe entre 40 et 45.

THE *SINANTHROPUS* CASTS

Excavations at Locality 1, Chou Kou Tien (Zhoukoudian or 'Dragon-bone Hill'), began in 1921-23. Among the excavated material sent for preparation to Uppsala, Sweden, were two hominid teeth (a lower left first premolar and an upper right third molar) discovered by the paleontologist Otto Zdansky.[1] This prompted, in 1927, the systematic excavation of the site under the direction of Birger Bohlin, with Davidson Black as acting director of the Cenozoic Research Laboratory. In that year another hominid tooth was found (a lower left first

---

1 Zdansky (1952) since has reported the discovery of a third hominid tooth in the breccia sent for preparation to Uppsala.

molar, no. 34 in Weidenreich 1937a), the specimen that led Davidson Black to establish the taxon *Sinanthropus pekinensis* (Black 1927). In December 1929, the first of the more complete skulls was found by W.C. Pei in Locus E (Skull III) (Black 1930a, 1930b). Following the death of Davidson Black in 1934, Franz Weidenreich was appointed director of the Cenozoic Research Laboratory.

Excavations continued until July 1937, when work at the site ceased because of the war with Japan. During the ten years of active excavation, a large number of hominid skeletal fragments were recovered from Locality 1, all now assigned to *Homo erectus pekinensis* (Campbell 1963). These included skulls, skull fragments, and facial bones from 13 individuals (not including maxillary fragment XIV from the Upper Cave [Weidenreich 1943: 16-17], 14 lower jaws, all but one probably from individuals different from those represented by the skull bones, 155 isolated or in situ teeth, seven femora fragments, two humeri fragments, one incomplete lunate bone of the wrist, a clavicle, and a fragment of an atlas.

Various parts of the Locality 1 site have yielded hominid specimens. These include the Main Deposit and the Lower Fissure, a cleft continuous with the Main Deposit but also extending below and to the north of it, and a separate part, the Kotzetang, some 20 m to the east of the Main Deposit. Eleven layers were recognized in the Main Deposit and Lower Fissure. For convenience, Black further identified the locations where hominid material was found as loci, giving them letters in alphabetical order as material was uncovered (Black et al. 1933). Later, excavation was carried on in 1 m squares and in 1 m depths termed levels. Thus, the fossils were originally identified on the basis of the locus in which they were discovered (e.g. the Locus E Skull or the Locus H mandible) and later also on the basis of the level at which they were found. Weidenreich (1943: 4-7) considered this system cumbersome because if more than one kind of fossil was recovered from the same horizon (e.g. the skulls from Locus L), then the identification became an involved one: Skull I, Locus L; Skull II, Locus L; Skull III, Locus L. He decided to number the cranial and extremity bones in numerical order, for example Skull 3 instead of the Locus E Skull, or Skull 12 instead of Skull III, Locus L.

In late November or early December 1941, the Chou Kou Tien hominids were packed for shipment to the United States for safekeeping. Sometime thereafter the entire assemblage of fossils collected to 1937, including the skeletons from the Upper Cave, was lost. Since that time, in spite of intensive searches by a number of people (cf. Shapiro 1974; Janus 1975), they have remained lost.

In 1949, three teeth were found in loose debris which had accumulated in the deposit since 1937. In 1951, two additional teeth plus a humerus and a tibia shaft were discovered in stored material excavated in 1937. Excavations resumed at Chou Kou Tien in 1958 under the direction of the Institute of Vertebrate Paleontology and Paleoanthropology and resulted in the discovery of a mandible in 1959 and, in 1966, the discovery of a premolar and additional parts of Skull v which was first excavated in 1934-36 (Howells 1977; Limbrey 1975; Chia 1975).

Because of the loss of the material excavated prior to World War II and the relative inaccessibility of the recently discovered fossils, studies that employ the Chou Kou Tien material rely on the Black and Weidenreich publications (Black 1927, 1929a, 1929b, 1930a, 1930b, 1931; Black et al. 1933; Weidenreich 1935, 1936, 1937a, 1941, 1943, 1948) and on the plaster casts produced by the University of Pennsylvania and the plastic casts offered by the Wenner-Gren Foundation. Unfortunately, neither the Pennsylvania nor the Wenner-Gren casts come from molds made directly on the original Chou Kou Tien fossils.

There are in existence, however, other casts of the Locality 1 Chou Kou Tien hominids. These come directly from molds made on the original fossils prior to their loss. Most of these primary casts are housed at the American Museum of Natural History in New York, but there are smaller collections at the British Museum (Natural History), in the G.H.R. von Koenigswald collections in Frankfurt, and at the Field Museum of Natural History in Chicago, the Harvard Peabody Museum, and the University of Pennsylvania Museum in Philadelphia. There may be original cast collections at other institutions, but I have not been able to trace them. I also have no knowledge of the number and kind of casts that remain in China, nor do I know if the original molds still exist.

Weidenreich (1943:iv), in his monograph on the *Sinanthropus* skull, records his appreciation of the technician who made the casts, Mr H Ch'en Chih, but this comment and a brief note by Black (1929b:209) on the future availability of casts of Skull III (Locus E) are all that have been written about them. Presumably there are still people in China who were directly associated with the molding and casting of the fossils, but I have not been successful in contacting them. I was able to contact Mrs Claire Taschdjian, who assisted Weidenreich in Peking from December 1940 until the outbreak of World War II in 1941. She stated that the molds were made by a technique that involves the fabrication of a large number of small plaster of paris pieces, each keyed to the others like a large three-dimensional jigsaw puzzle. The plaster pieces that have a side facing the fossil surface are polished with wax to prevent wet plaster from sticking to the mold when a cast is made. Each plaster piece in the mold has to fit precisely with the ones around it, otherwise the casts will be distorted. Molds made in this way, especially of large fossil pieces, are very intricate and quite large, but a skilled mold maker can produce casts from plaster molds which are nearly as accurate and detailed as those made today with silicon and latex molding rubbers.

TABLE 4.1
Location of primary casts of the Chou Kou Tien fossils and of the few remaining original specimens

LOCATION OF 'PRIMARY' CASTS
1. American Museum of Natural History, New York
   a. Department of Anthropology
      Skulls: I, II, III, V, VII, VIII, IX, X, XI, XII
      Adult mandibles: A II, G I, H I, M I
      Juvenile mandibles: B I, B III, B IV, C I, F I, B V
      Facial fragments: L I lt. maxilla & zygomatic, L II maxilla, L II(?) palate
      Teeth: One molar (34), M_1 lt., 6 incisors, 5 canines, 9 premolars, 12 molars
      Postcranial: Femora I-VII, Humeri I-II, clavicle, lunate, atlas

   b. Department of Vertebrate Paleontology
      Skulls: I, II, III, IV, V, VI, XII, XIII(O I maxilla)
      Adult mandibles: H IV, K I, M I, M II
      Juvenile mandibles: B I, B V
      Postcranial: Femora I-VII, Humeri I-II, clavicle, lunate, atlas
      Teeth: 12: 1 incisor, 3 canines, 3 premolars, 3 molars, 1 dpm_2 lt., 1 d$\underline{c}$ lt.

2. British Museum (Natural History), London
   Sub-Department of Anthropology
   Teeth: 35: 6 incisors, 5 canines, 9 premolars, 14 molars, 1 dpm_2 lt.
   (One of ten sets produced)

LOCATION OF ORIGINAL FOSSIL SPECIMENS
1. Palaeontological Institute, Uppsala, Sweden
   Department of Historical Geology & Palaeontology
   Teeth: P_3 lt., M^3 rt. (no Weidenreich catalog numbers)

2. Academia Sinica, Peking
   Institute of Vertebrate Paleontology & Paleoanthropology
   Skulls: Additional pieces of Skull V (1966)
   Adult mandible: 1959 mandible with M_1 lt.
   Teeth: C (P_3), I^1 lt., P^3rt., P^4rt., M_1lt., M_2 lt. (in material excavated pre-World War II)
   Postcranial: humerus, tibia (in material excavated pre-World War II)

Although most anthropologists concerned with studies of human evolution are aware of these original casts, what is not often appreciated is the number and high quality of primary *Sinanthropus* casts that are available. Table 4.1 lists the primary casts of *Sinanthropus* fossils and their locations. The table also includes a list of the few original specimens whose whereabouts are known. Table 4.2 lists those fossils for which I have been unable to locate casts. The *Sinanthropus* casts

TABLE 4.2
Chou Kou Tien fossil specimens with no known casts

Skull IX (J IV): 4 cranial fragments with connection (J IV frontal casts in American Museum of Natural History, Department of Anthropology & Department of Vertebrate Paleontology)
Adult mandible B II (condyle)
32 isolated teeth

---

available for study include all the mandibular fossils except the B2 condyle, all of the cranial pieces except the Skull IX (Locus J 4) fragments, all the postcranial bones, and all but 32 teeth.

The casts of the isolated teeth (listed by Weidenreich [1937a] catalog numbers in Table 4.3), 37 in all, are a representative sample of upper and lower incisors, canines, premolars and molars, and a lower second deciduous molar (no. 139'), which is the only deciduous tooth discussed by Weidenreich (1937a) that is not implanted in a mandible. I understand from Professor von Koenigswald that ten sets, all comprising the same 35 isolated teeth, were sent by Weidenreich to various institutions, but I have been able to trace only four sets and part of a fifth. The other two casts of teeth not represented in the sets deserve mention. A lower first molar (no. 34 in Weidenreich 1937a), not listed in Table 4.3, can be found only in the anthropology department of the American Museum of Natural History and was the specimen molded for the Wenner-Gren cast of this tooth (WGF no. O-CH 12). The other tooth, a left upper deciduous canine, presents an identification problem. In the monograph on the *Sinanthropus* dentition, Weidenreich makes no mention of a maxillary deciduous canine and, indeed, points out that no maxillary deciduous teeth were discovered (1937a:15). The only cast of this specimen is housed in the Department of Vertebrate Paleontology at the American Museum of Natural History (cat. no. 32737) with a catalog entry noting that the specimen comes from Locality 1, Chou Kou Tien, and was a gift of the Cenozoic Research Laboratory in 1940. I initially suspected that it belonged to one of the Upper Cave skeletons, but the almost completed root and little occlusal wear makes it difficult to relate to either the Upper Cave infant or the young child. In the absence of further identification, I am inclined to provisionally attribute the tooth to the Locality 1 sample.

All of the *Sinanthropus* primary casts are skillfully crafted and very beautifully painted. The amount of detail is remarkable, and comparison of the casts with published photographs of the original fossil specimens demonstrates that great care was taken to make the casts as representative of the originals as possible. Small cracks in the enamel of the teeth, and interstitial and occlusal wear facets can clearly be seen on the casts.

TABLE 4.3
Comparative measurements of available Chou Kou Tien isolated teeth casts

Catalog number	Weidenreich (1937a)		British Museum (Natural History)		Am. Mus. Nat. Hist. (Anthropology)		Am. Mus. Nat. Hist. (Vert. Paleon.)	
	B.-L.	M.-D.	B.-L.	M.-D.	B.-L.	M.-D.	B.-L.	M.-D.
2 (I^1 R)	7.9	9.8	—	9.8	7.6	10.0		
4 (I^1 L)	8.1	10.7	—	10.8	8.05	10.7	7.9	10.5
6 (I^2 R)	8.2	8.3	—	8.2	8.25	8.3		
135′ (I_1 R)	5.8	6.7	5.8	6.8	5.9	6.8		
8 (I_2 R)	7.1	7.2	—	7.1	6.6	7.25		
10 (I_2 L)	7.1	7.2	—	6.9	7.2	7.0		
13 ($\underline{C}$ L)	9.9	9.3	9.8	9.2	9.7	9.2	9.9	9.1
14 ($\underline{C}$ R)	10.6	9.6	10.5	9.5	10.5	9.5		
16 ($\underline{C}$ R)*	10.4	10.5	10.5	10.5	10.6	10.4	10.7	10.5
17 ($\overline{C}$ L)	10.4	9.0	8.8	8.8	10.0	8.6		
70 ($\overline{C}$ L)	—	8.4	8.2	7.9	8.3	8.2	8.25	8.1
19 (P^3 R)	12.9	9.2	12.9	9.2	12.8	9.1	12.9	9.15
25 (P^4 L)	11.3	7.9	11.5	8.1	11.4	8.0		
28 (P^4 R?)	10.3	7.2	10.1	7.1	10.25	7.1		
90 (P^4 R)	9.8	9.0	9.8	9.0	9.7	8.8		
20 (P_3 R)	10.0	9.3	10.6	9.2	10.2	9.25		
80 (P_3 R)	9.1	8.8	9.2	8.3	9.1	8.7	8.8	8.7
82 (P_3 R)	10.6	9.0	10.5	8.9	10.85	8.6		
89 (P_4 R)	8.0	9.9	6.9	9.9	7.1	9.7	7.1	9.9
29 (P_4 R)	11.1	9.2	10.8	9.2	10.9	8.9		
33 (M^1 L)	13.4	12.1	13.5	12.0	13.2	11.9		
140′ (M^1 L)	13.7	11.1	13.7	11.2	13.8	11.2	11.65	11.3
40 (M^2 L)	12.2	12.2	12.1	12.1	12.1	12.4		
41 (M^2 L)	13.2	11.1	13.2	11.0	12.8	10.9		
46 (M^3 R)	10.9	9.1	12.0	9.2	11.9	9.1		

Catalog number	Weidenreich (1937a)		British Museum (Natural History)		Am. Mus. Nat. Hist. (Anthropology)		Am. Mus. Nat. Hist. (Vert. Paleon.)	
	B.-L.	M.-D.	B.-L.	M.-D.	B.-L.	M.-D.	B.-L.	M.-D.
36 (M_1 L)	12.8	14.1	12.6	14.1	12.75	14.0	12.8	14.0
38 (M_1 L)	10.1	9.9	10.2	9.9	10.2	9.7		
137' (M_1/M_2 L)	11.7	13.0	11.5	13.0				
43 (M_2 L)	11.1	12.8	11.0	12.9	11.0	12.8		
44 (M_2 L)	11.5	13.1	11.5	13.2				
45 (M_2 R)	12.0	12.1	11.9	11.7	12.1	11.8		
51 (M_3 R)	12.1	12.2	11.5	12.1	12.1	11.6		
52 (M_3 R)	11.4	12.2	11.4	12.2	11.5	12.3		
131' (M_3 R)	10.8	12.7	10.9	12.7	10.8	12.7	10.8	12.9
139' (dpm_2 L)	8.9	10.9	9.2	11.0			9.3	10.9
undescribed dc L							6.1	7.5

* No. 16 may be a lower canine.

48 *Homo erectus*

To evaluate the accuracy of the casts, I measured a number of the cranial and mandibular specimens as well as the sets of isolated teeth at the British Museum (Natural History) and at the departments of Anthropology and Vertebrate Paleontology of the American Museum of Natural History. In Table 4.3, the dental measurements are compared with the published dimensions recorded in Weidenreich (1937a) and reveal that the casts are highly accurate. It is possible that the consistent differences in the buccolingual dimensions of the second lower premolar (no. 89) and the upper third molar (no. 46) between the casts and the published measurements may indicate typographical errors in the monograph; if the casts were inaccurate, a replica of this size would probably have inaccuracies in the mesiodistal dimensions as well.

The comprehensive collections of *Sinanthropus* primary casts at the American Museum of Natural History are divided between the departments of Vertebrate Paleontology and Anthropology, with the latter possessing the more complete set. Casts of some specimens can be found in both departments, and there is little difference between them; presumably, they were cast from the same molds. However, the anthropology department possesses a unique collection of hollow skull casts, replicas that preserve the entire endocranial area; those in vertebrate paleontology are of the better-known filled-in variety. Mrs Taschdjian recalls that these hollow casts were made at the Peking Union Medical College in 1941, after Weidenreich had left China, at a time when there were fears concerning a possible Japanese invasion. I have not located any additional hollow skull casts at other institutions; skull casts at the Harvard Peabody Museum, the University of Pennsylvania Museum and the British Museum all have filled-in braincases.

There is obviously no substitute for an examination of original fossil specimens. In the case of the Chou Kou Tien Locality 1 sample, however, this is impossible at present. The collections of primary casts at the American Museum of Natural History provide a practically complete and extraordinarily detailed and accurate set of replicas, and one can hope that this interesting series will be more fully utilized in the future.

THE PALEODEMOGRAPHY OF THE *SINANTHROPUS* SAMPLE

At various times during the course of excavations and later, after he had left China and begun work in New York, Weidenreich considered the number of individuals represented in the Locality 1 deposits and their age and sex distribution. As new specimens were recovered and the assignment of previously discovered fossils was reconsidered, different estimates for the number of individuals and their age and sex appeared in his published works. In 1935, Weidenreich recorded the number of individuals excavated up to that time (which did not include specimens from loci I to O) as 24 or 25, depending upon whether the Locus E skull (Skull III) was the same individual as the F1 mandible (Weidenreich 1935:449). In 1939, the count had reached 'approximately 38' (1948:195), and in 1941, in his last estimate, the number was given as 'about 45 individuals, adults and children, males and females' (1941:29). In this article, Weidenreich did not elaborate on the exact number of males and females or on the numbers of immature and mature specimens, but because in his various publications he provided an age estimate and sex identification for all the fossil specimens, except the atlas fragment, it is a simple matter to enumerate the age and sex groups. Table 4.4 is a comprehensive list of the individuals identified by Weidenreich, with a notation of the locus from which they were recovered, the age and sex assignments, and the specific skeletal and dental elements assigned to each individual by Weidenreich in his various publications. The total of 43 individuals requires additional comment. It does not include the upper deciduous canine discussed earlier, the lower premolar (P_4?) described by Zdansky (1952), or any of the fossils recovered from Chou Kou Tien after World War II (see Table 4.1), except the premolar assigned to the Locus H skull (Skull V). Nor does it reflect the possibility that some of the indicated individuals may belong together. For example, the atlas fragment, listed as individual I3, may belong with the I1 individual. Similarly, Weidenreich (1935: 447-8; 1943: 180) believed that it was possible that the E1 individual (Skull III) belonged with the F1 mandible, and that the H1 mandible and the H3 skull (Skull V) belonged to one person (1935:429). In an earlier publication, Weidenreich (1935) had referred to individuals C4 and D2; later, he indicated these individuals belonged with I1 and L1 respectively.

Table 4 does not include Skull XIV, the maxillary fragment from the Upper Cave. Weidenreich (1943:16-17) questioned its inclusion with the *Sinanthropus* fossils, but finally concluded that its similarities with the Skull XIII (Locus O) maxilla and its stage of fossilization warranted its placement with the *Sinanthropus* fossils. It is likely that Skull XIV is older than the Upper Cave *Homo sapiens* material but considerably younger than the rest of the *Sinanthropus* specimens, and therefore I have omitted it from the list. Finally, Limbrey (1975:59) refers to individual B6, composed of a radius fragment attributed to Locus B by Black. Weidenreich (1941:4), however, has identified this fossil specimen as probably belonging to a macaque.

In view of the above considerations, the total number of individuals recovered from Locality 1 may be as high as 45, if E1 and F1, H1 and H3, and I3 and I1 do not belong to the same individuals and if the recently discovered specimens belong to individuals not already represented in the count. Conversely, if these specimens do belong together, and if the new material goes with individuals

TABLE 4.4
Number of Chou Kou Tien Locality 1 individuals and associated bones*

Individual number	Locus	Individual	Sex/Age (F.W. assignment)	Composed of:
1	A	I	F 7-8	1 upper, 4 lower teeth
2	A	II	F Adult	1 upper, 3 lower teeth, mandible
3	A	III	M Adult	1 upper, 1 lower tooth
4	B	I	F 8-9	4 upper, 11 lower teeth, mandible
5	B	II	M Adult	2 upper, 1 lower tooth, mandible, Skull I, humerus, lunate
6	B	III	M 8-9	5 lower teeth, mandible
7	B	IV	F 5-6	1 upper, 6 lower teeth, mandible
8	B	V	M 11	9 lower teeth, mandible
9	C	I	F 8-9	1 upper, 2 lower teeth, mandible
10	C	II	M Adult	1 upper, 1 lower tooth, Femur I
11	C	III	M 9-10	4 lower teeth
12	D	I	F Adult	3 upper, 5 lower teeth, Skull II
13	E	I	M Juvenile	Skull III
14	F	I	M 8-9	3 lower teeth, mandible
15	F	II	M 5-6	3 upper teeth
16	F	III	F Adult	1 upper, 2 lower teeth
17	F	IV	M 13-14	4 upper teeth
18	G	I	M Adult	11 lower teeth, mandible
19	G	II	M Juvenile	Skull IV, clavicle
20	H	I	F Adult	2 lower teeth, mandible
21	H	II	F Adult	1 upper, 1 lower tooth
22	H	III	M Adult	Skull V, P_3
23	H	IV	F Adult	5 lower teeth, mandible
24	I	I	M Juvenile	2 upper, 2 lower teeth, Skull VI
25	I	II		Skull VII
26	I	III		atlas (same as I I??)
27	J	I	F Juvenile	Skull VIII

Individual number	Locus	Individual	Sex/Age (F.W. assignment)		Composed of:
28	J	II	F	Adult	Femur II
29	J	III	M	Adult	Femur III, humerus
30	J	IV	M	c.6	Skull IX
31	K	I	M	Adult	5 lower teeth, mandible
32	K	II	F	7-8	1 lower tooth
33	L	I	M	Adult	8 upper teeth, Skull X
34	L	II	F	Adult	13 upper teeth, Skull XI
35	L	III	M	Adult	Skull XII
36	L	IV	M	Juvenile	1 upper, 3 lower teeth
37	M	I	M	Adult	3 lower teeth, mandible, Femur VI
38	M	II	F	Adult	mandible
39	M	III	M	Adult	2 lower teeth, Femur VII
40	M	IV	M	Adult	Femora IV & V
41	N	I	F	4-5	1 upper, 1 lower tooth
42	O	I	F	Adult	6 upper teeth, Skull XIII
43	O	II	M	Adult	1 lower tooth

* The assignment of postcranial bones with cranial and dental material is, in most cases, indirect and should be viewed with caution.

TABLE 4.5
Age and sex distribution of the Chou Kou Tien Lower Cave specimens according to Weidenreich*

	Female	Male	Total
Immature	7 (39%) (39%)	11 (61%) (46%)	18 (43%)
Mature	11 (46%) (61%)	13 (54%) (54%)	24 (57%)
Total	18 (43%)	24 (57%)	42

* Percentages at right of figures refer to age distribution; those below the figures to sex distribution.

already accounted for, the total number of individuals may be 40. Therefore, a reasonable estimate of the number of individuals from Locality 1 is between 40 and 45.

The age and sex distribution of the Locality 1 fossils, based on the determinations of Weidenreich (1935, 1936, 1937a, 1943), is summarized in Table 4.5. The total of 42 individuals includes all the individuals listed in Table 4.4, omitting the atlas fragment I3. I have not included the fossil material recovered from Locality 1 since the end of World War II. Analysis of the cast collections at the American Museum of Natural History, in conjunction with Weidenreich's published reports, suggests that the age distribution is, with several exceptions, a reasonable one. The exceptions are the G2 individual, composed of Skull IV and a clavicle portion, and the L4 and E1 individuals. Skull IV, a fragment of a parietal bone found in three pieces, Weidenreich recorded as a young, possibly adolescent or young adult, male (1943:10). In discussing his reasons for this designation, Weidenreich (1943:179-81) compares the state of the sutures of Skull IV with those of Skulls III (Locus E) and XII (Skull III, Locus L), noting that all three exhibit open sutures without any hint of closure. Weidenreich considered Skull XII mature and Skull III immature, thus making it difficult to reach any conclusion concerning the age of the G2 individual.

The E1 skull was initially identified by Black (1929b:208) as an adolescent or early adult female, on the basis of its modelling and the delicacy of the features. Later, after preparation had revealed additional details of the bones, including exposure of the sutures, Black (1931: 25-7) modified his opinion, suggesting the E1 skull was an early adolescent male, equivalent in age to a modern twelve-year-old. Weidenreich (1943:180) agreed with Black, but assigned to the specimen a slightly younger age of eight to nine years. He based this conclusion on four points, including the presence of a cleft in the tympanic bone and the presence of a lacrimal groove, both of which he suggested were not present in adult individuals. Thirdly, he noted that the frontal torus was small, and finally he stated that the glenoid fossa was so small that only the condyle of the unquestionably immature F1 mandible fits. Weidenreich goes on to note that 'judged on the basis of the appearance of the skull of a child of modern man of this age, Skull III (E1) does not look like the skull of a child of eight or nine' (1943:181). Having examined the casts of this specimen at the American Museum, including the beautifully crafted set of casts of the individual cranial bones, I was impressed by the development of the muscle markings and the size of the bones. Whether this individual is a young male or a young adult female, however, is a difficult point to determine.

The L4 individual, composed of four teeth, is considered an adult by Weidenreich, although the stage of development of the lower third molar would suggest 10-12 years.

The sex determination of the Locality 1 specimens is more tenuous. Weidenreich based sex identification on the size variations present in the material, noting that 'In the Sinanthropus jaws and teeth, the extraordinary differences in size and thickness are conspicuous' and that 'the examination of the complete skeletal material, including the teeth, gives evidence that the differences in size of the teeth hold good for all the teeth, so that one can distinguish between a large and a small type' (1935:441). He concluded that the size differences in the teeth, and also in the cranial and postcranial bones, are related to sex differences. He further concluded that the large types were males and the small types, including the few whose measurements were intermediate, were females.

Perhaps the most reasonable way of examining Weidenreich's sex determinations of the Locality 1 material is to divide the sample into anatomical groups, looking first at the dentition, which Weidenreich considered extremely important in identifying the sex of the fossils, and then the skulls, mandibles, and postcranial specimens.

In dealing with the dentition, Weidenreich (1935: 442-3; 1937a: 131) considered the crown area (length breadth) to be an effective means of distinguishing sex differences. By assuming that the large teeth were those of males and the intermediate and smaller teeth were those of females, he divided the samples for each particular tooth into two groups, computed the means of crown areas for the large and small types of each tooth, and graphically represented these differences for the lower teeth (Weidenreich 1937a Atlas: 110-11). He provided no such data for the upper teeth, probably because the

54 Homo erectus

TABLE 4.6
Comparison of crown areas (in mm^2) of mandibular teeth taken directly from measurements provided in Weidenreich (1937a) with those graphically shown in the Weidenreich (1937a) Atlas

	Areas computed from measurements in Weidenreich (1937a)					Areas from 1937 Atlas		
	Females		Males		♀ to ♂ ratio	Female	Male	♀ to ♂ ratio
	N	M̄	N	M̄				
I$_1$	2	37.9	3	41	96.8	38	43	88.4
I$_2$	5	46.7	3	49.2	94.9	46	49	93.9
C	3	68.9	3	89.1	77.3	68	90	75.5
P$_3$	6	77.5	5	92.8	83.5	78	100	78
P$_4$	2	81.4	4	94.7	85.9	82	93	88.2
M$_1$	5	123.4	8	162.2	76.1	128	164.5	77.8
M$_2$	4	143.4	5	156.9	91.4	135	161	83.8
M$_3$	3	116.5	6	137.3	84.9	110	155	71

sample sizes, especially that of the large upper teeth he considered male, were very small. Weidenreich did not enumerate the individual teeth he employed in constructing the graph, but presumably they were drawn from the lists of teeth given in the text, where sex assignments and measurements were listed.

The absence of the specific data employed in reaching the graphic conclusion is unfortunate, because computation using the measurements of individual teeth listed by Weidenreich (1937a) results in mean crown areas in some cases significantly different from those provided in the graph (Table 4.6). In computing these figures, I checked the sex assignments of the teeth with the summary tables given in the dental monograph (Weidenreich 1937a: 7-14) and included the calculated crown areas of each tooth, except in the several cases where antimeres from the same individual were listed; in this situation, I averaged the individual length and breadth measurements of each tooth and used the area of these averaged dimensions as a single specimen. This has the advantage of minimizing unequal contributions of single individuals in the sample. I cannot explain many of the differences between the summary figures presented by Weidenreich and those computed from the individual dental measurements. In attempting to determine the samples from which Weidenreich calculated his mean areas, I manipulated the measurements of individual teeth, placing some specimens identified in the text as male in the female category if Weidenreich expressed doubts as to the sex assignments. For example, the mean crown areas of the lower third molars show a wide difference (Table 4.7, parts 1 and 2), and Weidenreich had questioned the designation of the D1 individual as

TABLE 4.7
Measurements by sex of the mandibular third molars from the Chou Kou Tien Lower Cave

1. Lower M3s. Measurements and sex according to Weidenreich (1937a)

	Females			Males			
	BL	MD	Area		BL	MD	Area
A2	10	10	100	G1	12.4	12.9	160.
H1	10.2	10	102	G1	12.3	12.	147.6
D1	12.1	12.2	147.6	AV	12.35	12.45	153.8
				B2	11.8	13.8	162.8
♀ area $\overline{M}$ = 116.3				F1	11.4	12.2	139.1
♂ area $\overline{M}$ = 137.3				L4	10.8	12.7	137.2
♀ to ♂ ratio = 84.9				M1	10.8	10.9	117.7
				M3	10.7	10.6	113.4

2. Crown areas in 1937 Atlas

♀ area $\overline{M}$ = 110
♂ area $\overline{M}$ = 155
♀ to ♂ ratio = 71

3. Crown areas if D1 M3 is subtracted from females and added to males; G1 teeth counted as 2 specimens

♀ mean area = 101.0
♂ mean area = 140.7
♀ to ♂ ratio = 71.8

female when discussing this specimen in the skull monograph (Weidenreich 1943: 178, 181) (although he did not question this determination in the dentition monograph [1937a] or elesewhere [1935:447]). The size of the D1 lower third molar (Table 4.7, part 1) is in the male range and its inclusion in the female group is a significant contribution to the high female mean. If the D1 molar is removed from the female group and placed with the males, and if the G1 molars are counted as two specimens and not averaged, then the female-to-male ratio, although not the male and female mean areas, is similar to that provided by Weidenreich (Table 4.7, part 3). However, this creates difficulties because Weidenreich had also questioned the male status of the L4 individual, and a consideration of all the teeth attributed to different individuals (see below) suggests that the M1 and M3 mandibles were also questionable. The means of the areas of other teeth are also difficult to interpret. The mean area of the male lower first premolars, for example, is given as 100 (Table 4.6), yet there are no lower first premolars whose area equals or exceeds 100. The largest is the left tooth from the G1 individual, whose area is 97.37 mm^2. It is clear that many questions can be directed towards Weidenreich's graphic representation of sex differences in the teeth, and these figures ought to be used cautiously.

Among the living primates, including humans, sexual dimorphism in the dentition is generally greatest in the canines, with the breadth dimension a better sex indicator than length (Wolpoff 1976). Unfortunately, sample sizes of mandibular and maxillary canines from Chou Kou Tien are very small, a total of six for each tooth. Frequency histograms plotted for length, breadth, and area of upper and lower canines are unimodal with the exception of the area of the lower canine crowns, but the small sample size for this tooth makes it difficult to interpret this result. The other teeth show a unimodal distribution.

What the frequency histograms illustrate is the very obvious fact that there are large teeth and small teeth, and many with intermediate dimensions. It was Weidenreich's contention that the large teeth are usually associated in large mandibles or with large, robust, thick-walled skulls and that the smaller teeth are associated in more gracile, thin-walled skulls and mandibles. To determine if this is really the case, I analyzed the associated teeth identified by Weidenreich (1937a) as belonging to the same individuals. Of the total sample size from Locality 1 of 40 to 45 individuals, 31 individuals possess one or more teeth but only 17 have four or more teeth. There are only one mandible and one maxilla that contain complete sets of teeth. The total of 31 individuals who are represented by teeth were identified by Weidenreich as 16 males and 15 females. Ten of these 16 males possess teeth whose dimensions are clearly larger than the mean values for the specific teeth, while 11 of the 15 females possess teeth smaller than the mean values for the specific teeth. The remaining ten individuals, B5, C1, F2, K2, L4, M1, M3, N1, O1, and O2, cannot, in my opinion, be accurately identified as to sex (Table 4.8). Indeed, Weidenreich had great difficulty in deciding the sex assignment for many of these.

In 1935, soon after it was found, Weidenreich described the juvenile jaw, B5, as male, 'according to the strength of the bones and the teeth' (Weidenreich 1935:431). In the 1937 dental monograph, after preparation had succeeded in removing the partially erupted first premolar and unerupted second premolar, he continued to refer to the specimen as male in the summary dental formula section, but noted after each of the B5 teeth that they may possibly be female; the dental measurements are intermediate. Similarly, the C1 individual, a mandibular fragment with three teeth, was identified as female, presumably because the breadth of the associated upper canine is somewhat smaller than the three canines he identified as male. The C1 mandible does possess, however, a second molar that is one of the largest in the series, both mesiodistally and buccolingually. Weidenreich (1937a:82) erroneously refers to the tooth (no.

TABLE 4.8
Identification of Chou Kou Tien individuals, using terms listed in Table 4.4

	Male	Female	Indeterminate	Total
Immature	B3, C3, F1	A1, B1, B4	B5, C1, F2, J1, J4, K2, N1	13
Mature	A3, C2, F4, G1, K1, L1, L3	A2, F3, H1, H2, H4, I1, L2, M2	B2, D1, H3, I2, J3, M1, M3, M4, O1, O2	25
Indeterminate	E1		G2, I3, J2, L4	5
Total	11	11	21	43

107) as male in his list of Chou Kou Tien lower second molars. The F2 specimen, composed of three isolated teeth, was identified as 'female (?)' in 1935, and as both male and female in various parts of the dental monograph (1937a). The teeth of this specimen are intermediate in size. The L4 individual, identified as a 'male (?)' by Weidenreich (1937a), consisting of one upper and three lower teeth, also possesses teeth intermediate in size. The lower molar associated with this individual, although categorized in the discussion as a first molar, is identified in the summary section as 'M1 (M2).' The relative lack of wear on this tooth, considering the stage of development of the associated third molar, suggests that it is a second molar; its dimensions would place it intermediate between large and small second molars.

The M1 and M3 individuals, both with several teeth and associated femoral fragments, are identified as males by Weidenreich, although the dimensions of their teeth are at the low end of the range, the third molars, for example (see Table 4.7), being far smaller than those of the G1 mandible. The O2 individual, represented by a single lower first molar whose dimensions are intermediate, and the K2 individual, represented by a single incisor, are simply too incomplete for any judgment. The N1 individual, composed of a lower second deciduous molar and an upper first permanent molar, was identified by Weidenreich as a female. However, the permanent molar is larger than other female teeth and approaches the size of the only male upper first molar in the sample.

Finally, the O1 maxillary fragment with six teeth, was judged by Weidenreich in the dental monograph to be female, presumably on the basis of the size of the incisor, premolars, and first molar, as Weidenreich did not identify any upper second or third molars as belonging to males. The first molar and second premolar are at the low end of the range, while sample sizes for the second incisor and first premolar are too small for any reasonable comparison. It is curious, then, that in the skull monograph Weidenreich (1943:16) notes 'as to sex, the size of the teeth indicate that it probably belonged to a male individual — male Individual O1.'

It seems apparent that a number of Locality 1 dental specimens cannot be adequately identified as to sex and that more variability exists in the sample than is usually recognized. This is also true of the cranial and mandibular material.

In the skull monograph, Weidenreich (1943:179) notes that for many of the specimens, 'the diagnosis is based on size or thickness, operating on the supposition that largeness and robustness of the bones is characteristic of males, and smallness and fragility of females.' But he also observes that sex differences in the skull are also 'suggested, in particular, by the character of the dentition' (1943:177). Examining the size of the teeth in the maxillae associated with skulls X (Skull I, Locus L) and XI (Skull II, Locus L), he concludes that the dentition and the size and robusticity of the cranial bones classifies these individuals as male and female respectively. But, he continues, 'such a classification encounters difficulties when the differences in size are not as pronounced as in these instances, or in cases where one must base one's classification on the calvaria alone, or on pieces of it, or even on fragments of individual bones' (Weidenreich 1943:178). Indeed, apart from skulls X and XI, sex identifications of the other cranial material are far more tentative. Skull XII (Skull III, Locus L), which shares many features in common with Skull X is probably male, as may also be the case of the young individual represented by Skull III (Locus E), although the determination of sex of this skull is hampered by its youth; Weidenreich ascribes the faint muscle markings on the nuchal torus and the relatively small supraorbital torus to young age rather than female gender. The remainder of the cranial material is composed of fragments of eight individuals, only three of which are associated with teeth.

The B2 individual, composed of Skull I, teeth, only one of which is complete enough for measurement, a mandibular condyle, Humerus I, and a lunate bone, is identified as male by Weidenreich. The measurable tooth, a lower third molar, is very large, but the fragmentary nature of the cranial material leaves the sex of this individual very much in doubt. Weidenreich (1936:11) questioned its status as male in the mandible monograph. The I1 individual, composed of four teeth and Skull VI (fragments of frontal, left temporal, and right parietal), Weidenreich judged to be female on the basis of small size and relative thinness of the cranial bone. The associated teeth are also small, suggesting that the identification of this individual as female is somewhat more secure than others. Individuals J1 and J4 are represented by fragments of the cranial vault, which, because of their size, are those of children. Although Weidenreich identifies the former as female and the latter as male, he expresses strong reservations about these assignments, and it seems relatively clear that the sex of these individuals is unknown. The G2 individual, composed of Skull IV, a small fragment of the right parietal, is identified by Weidenreich (1943:10) as male because of the thick bone and well-developed temporal lines. However, its bone thickness is significantly less than that of the female Skull XI, and the presence of well-marked temporal lines is not enough of an indicator to warrant determination of this skull as male.

The I2 individual, composed solely of the mastoid angle of a right parietal bone (Skull VII), is too incomplete for identification. The H3 individual, until the recent discovery of additional parts of the skull and an upper canine (Chia 1975), was composed of portions of both temporal bones (Skull V). Although Weidenreich (1943:10) argues in the skull monograph that the massive size of the bones identified this individual as male, in 1935, when considering the possibility that the H1 mandible and the H3 skull might belong to the same person, he noted that the combination would result in Skull V being designated as female (the same as the H1 mandible).

Finally, we must consider the D1 individual (Skull II and eight teeth). As noted earlier, Weidenreich had a great deal of trouble in assigning a sex to this individual. While many of the teeth are clearly in the small size range, some are very large, and the skull also possesses features of intermediate character. Black (1930b, 1931) had at first identified this specimen as male but later changed his mind. Weidenreich noted that the skull is larger than the female Skull XI, but he apparently did not believe it was large enough to be classified as male. He finally suggested:

There must have been, within the *Sinanthropus* population, a smaller type independent of sex. On the basis of existing materials I am unable to produce further evidence of this supposition, except for the fact that in the two groups of teeth — large and small — of which the large have been attributed to males and the small to females differences in size still remain which cannot be satisfactorily explained. It would not be surprising if, some day, we should discover dwarfs or at least a population distinctly smaller than those which we now consider the *Sinanthropus* type. (Weidenreich 1943:178.)

It would seem reasonable to suggest that Weidenreich was perhaps a little too concerned with the rigid division of his sample into large and small forms and that the difficulties he faced were prompted by a necessity to force a variable sample into differing types.

Similar observations can be directed to the analysis of the mandibular sample. In this case, Weidenreich (1936) was not dealing with the total sample when he wrote the monograph describing the mandibles, which contains discussions of 11 specimens, six immature and five mature, but without mention of the adult K1, M1 and M2 mandibles. In the monograph, Weidenreich had only three adult mandibles complete enough to study in detail, A2, G1, and H1. The G1 mandible is clearly much larger than the other two, and this feature combined with the very large teeth suggests male status. In many ways, the K1 mandible is similar to that of G1, so that it too may be identified as male. The A2 and H1 mandibles are much smaller in size than the previous two and possess teeth clearly in the low end of the range, suggesting that they may be considered females. However, Chia (1975) has recently suggested that the H3 individual, for which additional pieces of the vault have been discovered, may represent a considerably more evolved form than the other Chou Kou Tien specimens, and this may have a direct bearing on the identity of the H1 mandible. This will be discussed below.

The M1 and M2 jaws were not described by Weidenreich in the mandible monograph but were identified in the postcranial monograph (1941: 8-9) as male and female respectively. Three teeth associated with the M1 mandible are intermediate, suggesting that its male status is in some doubt, while the edentulous M2 jaw is similar in size to that of the A2 and H1 female specimens.

Of all the Locality 1 material identified as to sex, it is most difficult to adequately assess the postcranial material. It is fragmentary and incomplete, comprising fragments or splinters of seven femora, two humeri, a clavicle, a lunate bone, and an atlas. There are no complete bones for comparison, and although Weidenreich (1941) in many instances associated these specimens with cranial and dental remains, the certainty of these associations is open to serious question. Weidenreich identified each specimen as to sex, mainly on the basis of robustness and muscle markings, but the range of variation in the sample is uncertain, and it is not clear whether the only femur fragment to be identified as female, the J2 individual (Femur II), is mature or immature. As the other six femora specimens are designated male, along with all the other postcranial material, I have serious reservations as to the sex identification of the postcranial bones.

In summary, Table 4.9 presents the Chou Kou Tien Locality 1 sample, with those specimens most probably male, those female, and a sizable group whose sex identity is unknown. It seems clear that considerably more variation exists in the Locality 1 sample than first proposed by Weidenreich. This suggests that the Peking hominids do not neatly divide themselves into two well-marked groups. Thus, sexual dimorphism in this sample of *Homo erectus* was not as strongly marked as it apparently was in the australopithecines (Wolpoff 1976).

Many of the Chou Kou Tien specimens are very difficult to interpret, since they possess features reminiscent of both males and females. The hominid fossil

TABLE 4.9
Summary of the age and sex distribution of the Chou Kou Tien sample, taken directly from Table 4.8

	Male	Female	Indeterminate	Totals
Immature	3	3	7	13 (20.2%)
Mature	7	8	10	25 (58.1%)
Indeterminate	1		4	5 (11.6%)
Totals	11 (25.6%)	11 (25.6%)	21 (48.8%)	43

record is notably fragmentary and incomplete; rarely are there found enough bones from a single individual to assign sex with a fair degree of assurance. Reconstructions of the extent of sexual dimorphism in extinct hominids (e.g. Brace 1973, Wolpoff 1976) based on the examination of fragmentary and incomplete fossils are, thus, hampered by the paucity of the evidence. In addition, hominid fossil samples are not biological populations, but derive from a considerable (but unknown) number of generations. Intergenerational variation in sex related features may well have occurred, resulting in difficulties in sex assignment when these temporally different individuals are lumped together into one sample.

The recent announcement of the discovery of additional parts of the H3 individual (Skull V) (Chia 1975, Chiu et al. 1973) reinforces this observation. Weidenreich (1943), on the basis of the material discovered in 1934, had suggested that this skull was probably the largest in the sample, with a cranial capacity estimated to be around 1300 cm^3, an idea confirmed by the newly recovered portions of the skull. Chia (1975) has interpreted this as suggesting evolutionary change between the rest of the Chou Kou Tien sample and the Locus H material from the Upper Travertine which appears to be, geologically and faunally, of a later age. Thus, it is possible that the H1 mandible, which more closely resembles the European Neanderthal sample than the G1 mandible (Weidenreich 1936), may represent a later hominid rather than a female contemporary of the G1 mandible.

It may be that our understanding of the degree of sexual dimorphism in earlier hominid populations will remain unclear until nonmetric tests, such as that suggested by Kiszely (1974) using citrate concentrations, are able to accurately and unambiguously identify male and female.

ACKNOWLEDGMENTS

I wish to express my appreciation to the departments of Anthropology and Vertebrate Paleontology at the American Museum of Natural History for allowing me to examine the Chou Kou Tien casts in their collections and, in particular, to Dr Harry Shapiro and Ms Priscilla Ward for their time and interest. I would also like to thank the staff of the Sub-Department of Anthropology at the British Museum (Natural History) for their permission to examine Chou Kou Tien casts. Mrs Claire Taschdjian provided me with much interesting information concerning the molding techniques in use in Peking before World War II. Professor Phillip Tobias and Professor G.H.R. von Koenigswald kindly provided me with advice and information about the Chou Kou Tien casts.

5

W.W. HOWELLS

Homo erectus in human descent: ideas and problems

Cette étude vise à examiner les caractéristiques d'*Homo erectus* et son utilité future, en tant que concept et catégorie taxonomique, dans l'étude de l'évolution de l'homme. Le concept d'*Homo erectus* y est étudié en fonction du matériel fossile et des constructions théoriques qui ont été et qui sont susceptibles d'être utilisées en vue d'en comprendre les caractéristiques, les origines, et l'évolution qui a abouti à *Homo sapiens*. L'ouvrage discute aussi de quelques indications actuelles qui mettent en doute l'existence d'une lignée purement africaine du genre *Homo*. En ce qui concerne l'évaluation des schèmes d'interprétation, *Homo erectus* semble occuper une place centrale; c'est en se basant sur lui que l'on mesure les fossiles hominiens des périodes antérieures, contemporaines, et postérieures. Sa véritable nature doit, cependant, être cernée avec plus de précision. On ne doit pas en faire un concept commode pour les besoins de la classification, ni y voir simplement une étape dans l'évolution qui a abouti à l'homme moderne. *Homo erectus* était un animal, mais il a fait ses preuves.

THE RECOGNITION OF *HOMO ERECTUS*

As Voltaire might have said, if *Homo erectus* did not exist it would be necessary to invent him. And, of course, a century ago Ernst Haeckel did just that; evolutionary logic then as now demanded a grade of humanity antecedent to and more primitive than either modern man or Neanderthal Man. Haeckel, Darwin's enthusiastic disciple, needed this grade for his human phylogeny, and so he made it up out of whole cloth and named it *Pithecanthropus alalus*, the speechless ape-man. This was before Eugene Dubois began finding the actual fossils at Trinil in Java, which in 1894 he named *Pithecanthropus erectus*. Thus in these events there appeared in first guise the idea, the fossils, and the name of *Homo erectus*. Today, so much later, the problems before us are the meaning in human evolution of *Homo erectus* and the continuing definition and usefulness of such a concept and taxon.

64 *Homo erectus*

Plenty of controversy followed Dubois' discovery of *Pithecanthropus*, none of it to the point here and all of it due to the lack of other human fossils and the general state of knowledge. Only some Neanderthals of Europe, recognizably different from the Java cranium and from modern man, were then known. Other important fossils appeared in due course during the first half of the present century. Early ones were Mauer and Broken Hill, as single specimens. The first real series from a single locality was that of Peking Man, which became, and remains, fundamental in defining *Homo erectus*, due in so great measure to Davidson Black. He associated his department and resources with the excavations at Chou Kou Tien after Otto Zdansky, with a Swedish expedition, had recognized the first two teeth by 1926. When Birger Bohlin found the third and best in October, 1927, Black at once saw their importance as representing a new form of man and published a description the next month to serve as the definition of '*Sinanthropus pekinensis* Black and Zdansky.' Thus he generously associated the name of the first discoverer with his own, as the formal describer of the new genus and species. But it was Black who literally put Peking Man on the map as during the following year he travelled through North America and Europe with the precious 1927 tooth carried always on his person in a small, specially made receptacle,[1] continuing to raise both the interest and the money which assured continuance of the work at Chou Kou Tien and production of the now famous body of human material.[2]

1 This anecdote has several versions. According to Dora Hood (1964), a then-colleague of Black's, Dr Heinrich Neckles wrote in 1960 that he had advised Black to have a brass capsule made, with a screw closure and a ring on top so that he could wear it around his neck on a strong ribbon; a Chinese mechanic in the physiology laboratory made the capsule. But Hood adds that it was generally reported to have been worn on his watch chain. Roy Chapman Andrews, who knew Black in Peking at the time, said (1945) that the tooth was 'carried in an ingenious receptacle. It was literally chained to him.' He did not say how. According to von Koenigswald (1955) Black had a thick gold chain made in Peking with a small hollow figure to hang from it, into which the tooth fitted exactly. Professor von Koenigswald told Oakley (1975) that this was gold, in the shape of a small Buddha, and has said he believes he heard the story from J.L. Shellshear, then professor of anatomy at Hong Kong. Also, he does not credit a capsule of brass being hung on a gold watch chain. But the object was never found after Black's death and so the truth cannot be learned. KPO wonders (very private communication) whether the metal involved had a bearing on its disappearance.
2 The importance of this body of material cannot be overstressed. While Trinil 2, Dubois' original skull, is the holotype of *Homo erectus*, the Chou Kou Tien specimens as part of the hypodigm, though not necessarily a synchronous population, serve the purpose of indicating the degree and kind of variation in this hominid and thus provide a broader basis for defining *Homo erectus* (cf. Le Gros Clark 1964). Naturally, this is one fact which makes the loss of the Chou Kou Tien material so calamitous. It also reminds us

Peking Man was shortly followed by another important set from one locality, the Solo fossils of Java, and then by further occasional single finds such as Swanscombe in England and Steinheim in Germany. In addition, of course, there were now Neanderthals aplenty, as well as many anatonomically modern men of the late Pleistocene which should all have been called *Homo sapiens* but which were more usually given names at least specifically unique. It is only too well known how liberally separate generic names were commonly bestowed on earlier or anatomically different hominids.

This was one aspect of the interpretation, or rather the lack of interpretation, of relations among the few known fossil forms. Other aspects were the simple notions, then prevailing, as to lineages on the one hand and evolutionary grades on the other. Lineages and phylogenetic trees in this period emphasized distinctiveness, with a tendency for different fossil forms each to be given a branch of its own, and with modern man forming the apex without special reference to any other. On the other hand many writers, particularly German, early began to assemble known fossils into formal grades of advancement, or *Stufen*. (Haeckel had over twenty *Stufen* leading through the animal kingdom to man.) Perhaps the most usual terms used later for hominid grades were variants of Archanthropine (for such as Java Man), Palaeanthropine (for Neanderthals), and Neanthropine (for anatomical moderns). Only Franz Weidenreich, Davidson Black's successor in Peking, attempted (e.g. 1946) to reconcile these essentially vertical and horizontal views of phylogeny. Annoyed by trees which seemed to make extinct sidelines of every fossil man, he incorporated them all in his own tree — shaped *en espalier* (or what I once, in 1959, called candelabra) — which had both horizontal grades and vertical lines of regional parallel evolution.

that the other populations of fossil hominids are usually represented by one or a few generally fragmentary specimens, not by samples, and that the representativeness of such specimens is an assumption. For example, Rhodesian man, probably most usually termed *Homo sapiens rhodensiensis*, has been supposed to be represented cranially by the Saldana 1 skullcap, the Cave of Hearths right mandibular body, and, according to some, the Eyasi fragments, in addition to the type, Broken Hill 1 (all but one of these have been given from one to three taxonomic labels specifically or generically distinct from one another), and Broken Hill 2, the second maxillary part. This last is much less archaic than the corresponding part of Broken Hill 1, but has had little attention, except from Wells (1947). He felt no decision could be made as to whether it represented a large-jawed modern man or the same population as Broken Hill 1 (to which it has some resemblances). In the latter case, the gap between Rhodesian and modern man would be greatly reduced and marked variation demonstrated for the former. How representative, then, would Broken Hill 1 be of its population in general? There is also an analogy with the marked differences between Omo 1 and 2.

Thus from Haeckel through the first half of the twentieth century there had been a slow increase in the number of fossil hominids, including australopithecines, accompanied by rudimentary attempts to place them in clades or grades. Given the paucity of actual material these attempts were hardly more than hypothetical patterns of arrangement, especially since they were obfuscated by the thicket of generic and specific names then existing. The situation called for simplification and theoretical refinement and, obviously, for more fossils.

THE INVESTITURE OF *HOMO ERECTUS*

Ernst Mayr (1951), speaking in 1950, made the argument in detail that by zoological standards one could justify the inclusion of the whole array of fossil hominids then known in a single genus, *Homo*, with three successive species, *H. transvaalensis*[3] (for the australopithecines), *H. erectus*, and *H. sapiens*. He made various points. In the first place, while there are no standard criteria for defining a genus, the category nevertheless carries the useful implication of a group of species occupying the same adaptive plateau or zone. He noted, for example, that placing chimpanzee and gorilla in separate genera defeated the function of generic nomenclature and hid the basic likeness of the animals. Thus, even the australopithecines (as then understood) had no unequivocal claim to generic separation from *Homo* and lacked the typical simian characteristics. *Homo erectus*, as represented by the Java and Peking fossils, was sufficiently different to call for a species distinction from *Homo sapiens* but nothing more.

At the same time, Mayr said in effect that these three species must not be looked upon as rigid grades; the species as viewed by modern systematists is polytypic and multidimensional, comprising subspecies as natural divisions. Other primates, including the apes, have well-defined subspecies, but in modern man there is less such definition because of interbreeding, and distinct local populations have traditionally been referred to as races. Therefore, the question of whether subspecies or races are more appropriate is not clear. In earlier hominids, however, greater geographical variation was evident. Neanderthal Man posed a problem: should he be a separate species, or simply a subspecies of

---

3 Writers following this suggestion today use *Homo africanus*. Mayr at that time believed *africanus* to be preoccupied in *Homo* and used *transvaalensis* as the next species name available. Mayr later (1963) concluded not only that *Australopithecus* was a good genus, but separable because the much higher level of brain evolution in *Homo* determined a different adaptive niche. He also recognized the two species of *Australopithecus*, *africanus* and *robustus*, as sufficiently differentiated to make it a matter of taste as to whether one admits a second full genus, *Paranthropus*. Robinson (e.g. 1972a) has reasons, based on his understanding of phylogeny, for the same separation.

modern man? Mayr suggested that subspecies within *Homo sapiens* might be the following: modern man, Neanderthal Man, and a common ancestor of both. Mayr did not go into possible subspecies of *Homo erectus* except to note that the Chou Kou Tien fossils (as *Homo erectus pekinensis*) should be considered as no more than subspecifically distinct from the Javanese fossils. However, it was to be expected that temporal overlap occurred, some populations of *H. erectus* surviving when *H. sapiens* was already present. I cite all this from Mayr, as of 1950, to bring out ideas that are recurrent in views of human phylogeny.

Although Mayr's suggestions were not entirely new,[4] his presentation of them was timely and fruitful. They encouraged others to abandon the varied formal names of most fossils and to see the latter in a framework of looser grades of progress, or anagenetic species, in conformity with modern ideas of species structure and evolution. Le Gros Clark (1955) expressed similar opinions as to the unnecessary multiplication of genera and species. Although he preserved *Pithecanthropus* as a genus for the Javanese and Chinese fossils in the first edition of his book, he changed this to *Homo erectus* in the second (Le Gros Clark 1964). He also provided formal definitions for the genus *Homo* and for *Homo sapiens*, *Homo neanderthalensis*, and *Homo erectus* (*Pithecanthropus*). The last was based purely on the then-known Peking and early Java specimens (not including Solo). This is worth noting, because Le Gros Clark's appears to be the last generally recognized attempt to define *Homo erectus*, and his definition has been referred to, perhaps more implicitly than explicitly, since then.

During the next two decades, methods of dating advanced enormously and the pace of discovery of fossils picked up, including varied finds that might be assigned to *Homo erectus*, such as 'Telanthropus,' Ternifine, Olduvai Hominid 9, and Petralona. As to phylogenetic interpretation the most coherent and ambitious attempt was the structure of Weidenreich, later amplified by Coon (1962). Weidenreich died in 1948, before Mayr had suggested *Homo erectus* as a paleospecies, and Weidenreich used older terminology for grades (above the australopithecines, which were not hominids to him), such as Archanthropine, these grades being divided into several smaller steps (Weidenreich was something of a horizontal splitter). The emphasis on grades was evidently his main idea, but because he also envisaged geographically distinct lineages his scheme formed a

---

4 Weidenreich in 1940 had actually made some similar observations and threw out the suggestion that *Pithecanthropus erectus* and *Sinanthropus pekinensis* should instead be called *Homo erectus javanicus* and *H. e. pekinensis*, a kind of terminology coinciding exactly with later views. But Weidenreich did not adopt this himself, continuing to use previously given names that he insisted were mere labels. Simpson (1945) also thought all known hominids above the australopithecines could be accommodated in one genus, *Homo*.

grid (e.g. 1946), of which the 'candelabra' pattern is an abstraction. He qualified this in various ways, viewing all fossil hominids as members of one species, asserting that evolution was progressing similarly over a vast area, though always with a tendency to racial differentiation. But casting his general interpretation in the form of a grid may have made it seem more simplistic than he meant.

Weidenreich tended to let morphology rule apparent chronology, and he may have been more correct in some instances than the geological daters of the time. Also, he was rather cavalier with terminology, stating in several places that, while holding the view that all hominids were indeed one species, he would go on using the illegitimate generic names as labels only, for convenience. He was particularly vexing in his use of 'Neanderthal,' 'Neanderthalian,' and 'Neanderthaloid' interchangeably, without clear definition, for an intermediate grade of progress above *Homo erectus*. This is confusing for Broken Hill and Solo because in some papers he referred to both as Neanderthals but in others (1943, 1946) explicitly stated that Solo could not be classed with Neanderthals but could be 'defined as an enlarged *Pithecanthropus*' (Weidenreich 1946:40). He placed both forms at an intermediate level in his grid, with Solo in fact in his Archanthropine grade just above Trinil and Peking.

Coon tightened up Weidenreich's loose formulations. He accepted the subsequent division of hominids into *Homo erectus* and *Homo sapiens*, both as species and, explicitly, as grades. In the former he placed, in addition to the early Java and Peking skulls, those from Solo, Broken Hill, and upper Bed II at Olduvai (OH 9). His has been the most systematic attempt to quantify distinctions between *H. erectus* and *H. sapiens*. He found exclusive ranges in indices of sagittal curvature of bones of the vault and also in indices relating brain size to palate area or molar tooth size. He thus assigned Solo and Broken Hill to *H. erectus*, somewhat against the grain of thinking of the time — thinking which was probably influenced by the supposed late dates for these two at the time he wrote. However, recent multivariate analyses (Stringer 1974; Corruccini 1974) also associate these crania with accepted *H. erectus* forms. In a new study that morphologically reexamines the original Solo Skulls, Santa Luca (1976, 1977) has concluded that they definitely approximate most closely the Java and Peking forms and are clearly distinguishable from modern man.

Although Weidenreich's grid contained more framework than fossils (because of the paucity of fossils), his hypothesis was more systematic and thorough than those of others at and before his time. Because Coon (whose assembling and analysis of morphological data was also a major accomplishment) accepted and extended Weidenreich's thesis of regional lineages leading independently to local races, his pushing of Weidenreich's attempt to reconcile grade and clade led to a paradox. He proposed that five already distinct subspecies of *H. erectus* had separately evolved into five subspecies of *H. sapiens*.[5] This has been generally rejected by others on grounds of genetic and evolutionary theory, though without substitute schemes of lineages offered to take its place.[6] In fact, the net effect of writers like Coon, Le Gros Clark, and Mayr has probably been to influence others to see fossil hominids primarily in terms of grades, i.e. *Homo erectus* and *Homo sapiens*, in spite of Mayr's original qualifications.

In summary, until very recently *Homo erectus* has probably had, in the minds of many, a general character such as would be served by Le Gros Clark's 1955 definition based on the Far Eastern fossils: a brain size varying around a mean of 1000 cm^3, a low thick skull with massive supraorbital ridges and marked constriction of the frontal just behind, and a heavy chinless mandible, along with such details as a thickened and more horizontal tympanic bone, a sagittal ridge, etc. He also noted that *Homo erectus* limb bones were not distinguishable from those of *Homo sapiens*, but finds and detailed studies since that time would suggest just the reverse. Slight to marked anatomical distinctions from modern man may be recognized in all the sets of limb material that have been assigned to *Homo erectus*, with doubt being cast on the positive association of the original Trinil femur, from which *erectus* became the species name, with the skull cap. In any event, by 1970 it might be said that regardless of possibly earlier or later specimens, of those falling in time between about 1.4 and 0.4 million years ago, none recognizable as *Homo* were known that would not be placed by anthropological concensus in *Homo erectus*.

PRESENT QUERIES

Over two decades *Homo erectus* became not only the accepted taxon but an accepted concept. In academic terms everyone was converted. '*Homo erectus*' is what students learn about, and we are probably at the precarious point where it is not only popular writers who may refer to *Homo erectus* as 'he' or 'him.' Are we in danger of a degree of semantic entrapment? It may be time to ask how much is taxon and how much is concept. What does '*Homo erectus*' actually represent to us? And, what should it represent to us today?

---

5 This is not a necessary paradox. See Simpson (1961) for the relationships of grades and clades over time.
6 Le Gros Clark could write in 1964 that the inference that *Homo erectus* was ancestral to *Homo sapiens*, though reasonable, 'must be accepted for the present as not very much more than a working hypothesis.' Given the evidence, this is not as conservative a statement as it sounds.

Put otherwise, the very currency[7] of the nomen has probably fostered the tendency to view hominids in terms of grades, ignoring cladistic distinctions. Again, it may be time to begin revising this once praiseworthy simplification in the light of new complexities arising not only from recent discoveries but also from a priori considerations of a theoretical nature induced partly by paleontology and zoology. It should be possible to improve the whole hypothetical framework of Middle Pleistocene developments relative to its present rather amorphous state, if only by applying some queries.[8]

Homo erectus is a paleospecies (i.e. in an evolutionary succession), not a biospecies of which the population genetic structure might be studied. As Mayr said long ago, the latter is polytypic and multidimensional except as to time, although it might have identifiable extinct subspecies (he suggested the Neanderthals in the case of our own species). A paleontological species, however, must also be expected to have subspecies, temporal as well as geographical. *Homo erectus* must have arisen from a parental species and by evolutionary change have given way to a successor species, *Homo sapiens*, in both cases by forms of speciation and replacement normal in animal evolution.

PATTERNS OF SPECIES SUCCESSION

Anthropologists see the process from two generally different viewpoints, each based on some loosely linked ideas as to homogeneity, rate of evolution, etc. Again, Mayr and others wrote about this some years ago, and the distinction has recently been formalized by Eldredge and Gould (1972) although the ideas are familiar. These writers oppose 'phyletic gradualism' to 'punctuated equilibria' as the patterns of species change.

In 'phyletic gradualism,' change is viewed as gradual and general over the species, so that in the case of *Homo erectus* the species as now recognized might turn out, on full information, to be no more than an arbitrarily limited span within an evolving continuum covering the whole phylum or species. In 'punctuated equilibria,' the apparent discontinuity, seen so often in a paleontological succession, is not simply the artifact of a gap in the record but is real.

7 I am all too aware how this may be fostered. When one is asked to write a 'popular' exposition of *Homo erectus* (Howells 1966), the need to be noncontroversial and intelligible leads to a sort of tidy and artistic simplification of the object. And the availability of such writing for reprinting and citing spreads the cut-and-dried form more widely.
8 Pilbeam (1975) has provided a recent overview of the material and the geological background, raising many of the problems and questions dealt with here and suggesting some solutions. This article should be consulted.

The process of change is not species-wide but results from allopatric speciation. A subspecies, ideally a peripheral isolate of the old species, becomes the new form in some significant respect and replaces populations of the old by migration. Thus the main body of the species does not undergo the gradual change to a new species. Other subspecies, of course, may also have changed, but without an evolutionarily successful result (unless by giving rise to another new and separate species).

Embedded in the above distinction is the simpler one already noted of grades and clades. As applied to *Homo erectus* it is evident that the first pattern, gradualism, with an accent on advance through evolutionary grades, has been the ruling interpretation. It is also known as unilinear evolution. This would cover the Weidenreich-Coon hypothesis, at least in Coon's version. Although recognition of clades or different lines of evolution is essential to their scheme, the differentiation is set at the base of *Homo* and is not progressive, not a tendency toward continuously increasing local divergence which might sometimes have led to replacement (although in fact both men did regard the Neanderthals of Europe as a dead end).

The point has often been made that in hominids the adaptive niche had become so broad that allopatric or adaptive speciation was not likely to occur. There is also argument as to whether human mobility and gene flow in the Pleistocene would have overcome any significant differentiation. Weidenreich, giving this factor some weight and stressing unity in diversity, early held that no more than one species of hominid (which for him did not include the australopithecines) had existed at one time period during the Pleistocene (cf. Dobzhansky 1944). In a broad sense this dictum had been adhered to by most writers (and is of course reflected in the general recognition of only *H. erectus* and *H. sapiens* from the Middle Pleistocene on). But perhaps this has not meant the same thing to all people.

It is the opposite view, of the contribution of diverging or cladistic evolution and eventual replacement by one segment (typically a subspecies) of all other segments of a species, that needs more consideration. It appears to be the only acceptable explanation of the Neanderthal-to-modern transition in Europe, though some argue warmly against it. Wider applications, as to *Homo erectus*, must be attempted but have been difficult, and not much has been done to establish local differentiation at a given time within *Homo erectus* in a formal way. Campbell (1965, 1972) has supplied appropriate trinominals for subspecies, but simply as a matter of correct taxonomic nomenclature for recognized finds. This cannot be said actually to have established subspecies. Stewart (1970) has made a general comparison of some mandibular traits, finding a distinction of western specimens (Ternifine, Mauer) from eastern ones in such things as greater interforaminal breadth, more prominent marginal torus, etc., traits that seem to foreshadow Neanderthals. Such attempts and some beginning multivariate analyses are about the limit for efforts to identify clades. (Weidenreich's 1943 study of the Java and Peking skulls, and covering Solo, Broken Hill, and the Neanderthals to a degree, is still the classic of this kind but hardly answers the main problems of differentiation.)

SPECIFIC PROBLEMS

Homo erectus, however the species is viewed, had an origin, a duration or ascendancy, and a replacement. Recent dates indicate, I suggested earlier, that over a million years, from perhaps 1.4 million to 400,000 years ago, all known specimens of *Homo* would be generally recognized as of *erectus* grade (to adopt for the moment a 'grade' point of view). This implies no more than it says, as there are doubtless earlier and later such specimens, contemporaneous with other hominids in both instances. But it marks off three problem areas where new facts should be beginning to lead to more positive hypotheses: (1) the circumstances of the emergence of *Homo erectus* and its relation to other hominids; (2) the mode of transition from *H. erectus* to *H. sapiens*; and (3) the degree and the pace of change in *Homo erectus* from beginning to end. I have no intention of attempting answers but only of pointing to problems and suggesting how new facts may be put into them. Let us start with the last.

Changes within Homo erectus
Was the rate of change steady and gradual (whether linear or logarithmic in scale), or did it show a period of stasis between epochs of more rapid progress, like a stair tread between two risers, in conformity to the pattern of 'punctuated equilibria'?

In general, writers have looked for and found evidence of continuous advance locally. Java is the best case, where such evidence has been perceived by all who have worked with the materials. Weidenreich (1946) had no less than four levels, including *Meganthropus* along with *Pithecanthropus robustus*, *P. erectus*, and *P. soloensis*, all four being successive stages within his Archanthropine grade. These have for long been allocated to three faunal zones: the first two forms in the Djetis, *P. erectus* in the Trinil, and *P. soloensis* in the Ngandong. In the emerging but still labile framework of Javanese dates, the Djetis forms have an age of 1.9±0.4 million years, the Trinil between 0.9 million years and 0.7 million years, and the Ngandong about 200,000 years or more (revised from the former estimate of much greater recency).

While not coinciding fully with Weidenreich or one another, or even finding neat successive levels, von Koenigswald (1975), Jacob (1975a), and Sartono

(1975) each also see at least three groups or types. The pattern is affected by new finds, particularly by two crania reportedly from well down in the Kabuh beds (Trinil fauna), Sangiran 17 or *Pithecanthropus* VIII, and the Sambungmachan specimen, which are so similar to the Ngandong Solo skulls that they are classed with them by Jacob.[9] He sees three groups: the early robust and smaller-brained *Pithecanthropus modjokertensis* of the Djetis zone followed by its Trinil-Ngandong descendent, *P. soloensis*, and by a contemporary and sympatric, relatively gracile *P. erectus*. Jacob (1975a, 1975b) makes detailed distinctions among the groups, finding for *P. soloensis* not only a robust face in addition to the heavy cranial traits already known from Ngandong, and suggesting (from the tibiae) a stature of 170 to 180 cm and a body weight of 60 to 100 kg. (He recognizes that others would put all these taxa in *Homo*, perhaps as subspecies.) Now, although the material has considerably expanded in recent years, it probably does not yet support any detailed phyletic scheme. The suggested existence of two distinct sympatric subspecies in Java appears puzzling if not downright dubious.

The above groupings have indeed all been put in *Homo* by other writers, so that they would appear as subspecies, seemingly chronologically successive. It has previously been accepted that the cruder Puchangan (Djetis zone) form (represented by few fossils) progressed locally to give rise to the later Trinil and Solo groups, which Jacob is suggesting to be contemporaries. But this traditional explanation of succession is not known to be the true one, and no explicit primary pattern of progress, or its rate, appears from the material or the opinions of those working with it; i.e. progressive stages are seen, possibly as the local 'gradualism' assumed by Weidenreich and Coon. Perhaps the most striking fact, if dates are approximately correct, is the opposite of gradualism: the long continuance – half a million years – of morphology from Sambungmachan to Solo without evident change. I will return later to the peripheral position of Java as possibly bearing on replacement.

In China, more briefly, Weidenreich saw no evolutionary trend in the Locality 1 hominids at Chou Kou Tien. Chinese paleoanthropologists definitely do, however, in a cranium restored from fragments found in 1966 and associated with a late level in the deposit. The cranial volume is estimated as about 1140 cm^3, higher than any previous reliable specimen, and they note other signs of advanced status. Also, the Lantian remains are placed substantially earlier in time, and are certainly lower in the scale of hominid development, having a

---

9 Santa Luca (personal communication) disagrees, believing Sangiran 17 to be at best transitional. Sartono (1975) places this specimen later in the Middle Pleistocene, i.e. closer in time to the supposed position of Solo, which would accord with a transitional status.

cranial volume and other aspects suggesting the Sangiran 4 (*Pithecanthropus* IV) parts from the Puchangan beds (Djetis zone) in Java. This all points to evolutionary advance in North China over time. On the other hand the two Yuan-mou incisors, admittedly rather narrow evidence, have been rated by the Chinese as morphologically equivalent to those of Locality 1, while being earlier in time. Since the conference (see addendum) Chinese scientists have developed paleomagnetic dates for Yuan-mou, giving an estimate of about 1.7 million years for the tooth-bearing stratum, or about a million years earlier than Locality 1 at Chou Kou Tien.

East Africa has put the matter in sharper focus with the excellently preserved and restored ER 3733 skull. It is recognized by Leakey and Walker (1976) as *Homo erectus* not by definition or assumed grade but on the sound ground of resemblance (not identity) to Peking Man. The specimen is at least one million years older than Locality 1. Furthermore it is succeeded in East Africa, at about 1.2 million years, by the very robust, large-browed Olduvai Hominid 9, the relative robusticity inferentially being supported by the innominate and femur of Olduvai Hominid 28.

So there is a dilemma. On the face of it a stage had been reached in East Africa equivalent to that present at Chou Kou Tien at least a million years later, whether or not we can entertain the possibility that such a population covered the time and the distance unchanged. If so, this and the Solo population would argue that some hominid groups could have remained constant in grade for very long periods. It does not argue that this would have been the same grade or level everywhere. The occurrence in the same areas of different or changed forms (e.g. OH 9 in East Africa) might indicate changes of pace but also capacity for considerable differentiation by clade within what we now call *Homo erectus*.

Replacement by Homo sapiens

Here we are bothered by fuzzy language. By present convention this species includes the Neanderthals, while on the other hand no formal diagnosis or definition for *Homo sapiens* exists, but only a sort of unrocked boat of agreement that any fossil apparently more advanced than *H. erectus* is *H. sapiens*.[10] The substantive matter is the timing and pattern of transition. By present signs this began about 400,000 years ago, with only Neanderthals and modern men certainly known after about 100,000 BP, so that the change would have been moderately rapid compared to the duration of populations of *H.*

---

10 Some writers would still meet the problem by using a chronological reference and by broadening the term 'Neanderthal' to cover any later Pleistocene hominids including Broken Hill and Solo.

erectus. During this phase there was probably a long overlap between the latter populations and some which were more advanced. A late 'primitive' might be the Salé skull from Morocco, with its small brain and estimated date of 165,000 years, while early 'moderns' would be the Omo skulls from the Kibish formation, for which a date of 130,000 years has been entertained.

As to pattern, the process ended, we know, with the survival of anatomically modern man alone by 35,000 BC. At that time populations having the locally appropriate racial characters of today were present in such widely separate regions as Europe, South Africa(?), and Australia (and quite possibly America). But these and *Homo erectus* do not meet each other halfway; there are no visible lines of connection across the time interval. For example, the modern-leaning Omo skulls do not, in multivariate analyses, preferentially approach African skulls among modern populations, but quite the reverse.

Suppose we argue that the classic Neanderthals were the culmination of one lineage deriving from *Homo erectus*. Steinheim, Swanscombe, and Fontéchevade, intermediate chronologically, have in the past also been viewed as transitional in form, though generally assigned to *Homo sapiens*. They have also been variously termed Preneanderthal or Presapiens, with the implications that they were either parental to both later populations or else possibly an early *H. sapiens* stock leading to modern man, which relinquished Europe to Neanderthals for a while only to return in developed form in the Upper Paleolithic – an awkward hypothesis. These earlier fossils have now been recently supplemented by various others (Arago, La Chaise caves, Cova Negra, etc.), and the de Lumleys and others prefer to call them all by the noncommittal term Anteneanderthals.[11] The great height and forward extension of the classic Neanderthal face, the essential cranial character, is not visible, but certain other Neanderthal traits are present (de Lumley and de Lumley 1975); they seem intermediate to Neanderthals and *H. erectus* in their mixture of features.

Steinheim and Swanscombe now appear, in multivariate analyses (Weiner and Campbell 1964; Stringer 1974; Corruccini 1974), to have closer relations with early and classic Neanderthals than most writers have heretofore allowed and do not suggest connections with any modern populations (Coon considered them to be early Caucasoids). Fontéchevade (Corruccini 1975; Trinkaus 1973) also appears as more Neanderthal than modern, contrary to former opinion. At present, then, it would seem that the Anteneanderthals are in fact Preneander-

---

[11] I am omitting supposedly pre-Holstein fossils which might be included by others, especially Vértesszöllös, Petralona, and Mauer. This is arbitrary, and done partly to be conservative as to time period and partly because the statements as to morphological intermediacy seem most safely made for the later fossils.

thal in the simple sense (see also Thoma 1976b; Piveteau 1976b). The classic Neanderthals of the late Eem and early Würm periods had to issue normally from some parental source, and local evolution and individualization in the preexisting population in the west looks increasingly like the logical explanation. (There is really no reason save a fixation on Europe to see these Anteneanderthals as giving rise to other late forms of *H. sapiens*.)

This would give us one diverging and evolving lineage arising from *Homo erectus* ancestors in the west over several hundred thousand years before being overrun and replaced by populations of a second well-differentiated lineage or clade, that of anatomically modern *Homo sapiens*. We do not know, of course, the original locale of the latter, but the above is the simplest hypothesis and surely the one to defend or defeat. (Of course, stated in so simple and brief a form, defending the hypothesis faces certain problems. There are the Neanderthals of Asia Minor including Shanidar, Tabun, and Amud, who must be viewed as part of the same evolution, and the still more easterly Neanderthals or approximations thereto, including Teshik Tash and Mapa, although Santa Luca finds Mapa more like an advanced Pekingese. In addition there is the confused Near Eastern situation created by Skhul and Qafza, which are persistently recognized by some as honorary Neanderthals. But I am hypothesizing in relation to the main theme, not trying to develop a thesis.)

A third clade could be one in eastern and southern Africa comprising the Rhodesian Man population (the Broken Hill individuals, Saldanha, and Cave of Hearths) and possible kin. Enlarging its representation and lineage at present is difficult because prospective members are few and isolated, and in this region we are especially bothered by sliding dates. We have for inspection ER 3733 at before 1.5 million years, a Peking Man-like form, and the quite different Olduvai Hominid 9 at about 1.2 million years with its flat skull and bulbous brows, then the Ndutu cranium (with a possible late Acheulean, i.e. Earlier Stone Age, association [Mturi 1976] suggesting contemporaneity with Saldanha), again resembling an advanced Peking Man (Clarke 1976) even in multivariate analysis, but differing considerably from Broken Hill 1. Only then do we arrive at the latter and, apparently in a similar time zone, Omo 1 and 2.

Omo 1, a robust modern, and even the more archaic Omo 2, would not superficially appear to be on the same evolutionary track as Broken Hill 1. Would this mean that Broken Hill 1 was nontransitional and without issue as has been suggested before (Pilbeam 1972)? Day (1972), however, does find some resemblance between Omo 2 and Broken Hill. So does Rightmire (1976) on the basis of a new close examination, finding sufficient similarity to allow possible derivation of Broken Hill from OH 9 (which is of course vastly earlier) as well as possible relatedness with later hominids like Florisbad in addition to Omo 2, though he is cautious. At any rate, rejecting any direct kinship with Neanderthals proper, Rightmire feels that a formal designation for Broken Hill, Omo 2, and Saldanha, at least, as *Homo sapiens rhodesiensis*, now seems justified and suggests that 'the populations sampled at Broken Hill and in the Omo are probably evolved from local groups of early Mid-Pleistocene *Homo erectus*.'

Are we converging on Coon here, with different subspecies of *Homo sapiens* developing separately from different subspecies of *Homo erectus*? The distinctions would be: (1) Rightmire (1976) is not saying that local *Homo erectus* populations were already subspecifically distinct in their regions (and ER 3733 might be called to testify the opposite), only that the descendant line here was divergent from that of Neanderthals to the north, and (2) he is not explicitly tracing a separate subspecies (race) of living man from a unique late Pleistocene ancestor (here *H. s. rhodesiensis*) as Coon did. But these are the kinds of propositions we are faced with (and which are cast up by such terminology as *H. s. neanderthalensis* and *H. s. rhodesiensis*).

There does not seem to be ground enough to hypothesize further afield, i.e. as to other clades, or what the Solo population might have led to, if anything. One argument for a single Mid-Pleistocene clade, alone ancestral to living man (whether *H. s. rhodesiensis* or another), that is, for the separate development and ultimate ascendancy of the anatomically modern line, is the cranial homogeneity of living races compared to other forms. In the expansion of this form, hybridization with others must be suspected and can be argued for from such evidence as the Skhul group or even Předmost. But the more one allows that some racial differentiation in modern man is to be ascribed to exogenous genes, the more one must admit an original homogeneity in the basic population.

All the above — the divergence of at least three possible lineages and the ultimate survival, due to some functional advantage, of only one of them — would conform to the ideal pattern of grades and clades (e.g. Simpson 1961:127) in which clades may eventually constitute separate grades. The pattern differs from the parallel progress of 'clades' in the Weidenreich-Coon hypothesis in that their scheme does not envisage replacement but only a modified phyletic gradualism.

Origins of Homo erectus

If *Homo erectus* is a species, not simply a morphological grade, then whether or not it branched later on it must have had a single origin. Here definition becomes a problem of importance. At the 1976 Nice Congress it was suggested that, sticking to accepted definitions, no example of *Homo* was known before 2 million years ago (e.g. at Olduvai, Omo, East Rudolph, Swartkrans, Sterkfontein [new find of Tobias] and, one can add, Sangiran). This brought a rejoinder that we should not have temporal rubicons. But a main problem is lack of definitions of *Homo*, above all, whether *Homo* other than *H. erectus* can be recognized in this time zone.

This involves us with the australopithecines. A thumbnail history will help with perspective. As is well known, two forms, gracile and robust, were discerned in South Africa. Interpretations have ranged from mere dimorphism in a single species through two species of *Australopithecus* (the centrist position) to Robinson's view (e.g. 1972a) of two genera sufficiently distinct adaptively to set *Paranthropus* aside and sink *Australopithecus* into *Homo africanus*, the direct ancestor of later species of man. In any case, for some time it has been generally agreed that *Homo* arose from australopithecines, a view which has carried on through the subsequent finding of hominids in East Africa.[12] Here the robust line (however called) has been found widely, coming down to about one million years and known to be contemporary with other hominids; it is confidently ruled out of direct human ancestry. It is with 'graciles' that confusion lies, the main argument having been whether there existed a gracile australopithecine in East Africa in addition to *Homo* and the robust form. *Homo habilis* as a species was created in 1964 (Leakey et al.) from finds in beds I and II at Olduvai, and its creators relaxed Le Gros Clark's 1955 definition of the genus *Homo* to accommodate it. Shortly, however, the Bed II specimens were placed by many in *Homo erectus* instead. Nevertheless, *Homo habilis* gained currency, with some agreement that the form constituted a bridge between the South African graciles and *Homo erectus* as represented by Olduvai Hominid 9 of upper Bed II.[13] Others have felt all along that *H. habilis* of Bed I was actually a gracile australopithecine advanced beyond the Sterkfontein form, for example, but not to be included in *Homo*. This view has recently been joined by Mary Leakey and is probably the prevailing one of the moment.

Other Early Hominines: Homo unexpectus
This shift is partly due to the new finds, which have upset the simpler picture, insistently suggesting the early presence of more developed forms of *Homo*, that is, before OH 9 and probably before the specimens of lower Bed II, which have been called *H. erectus*. A particularly surprising aspect has been the apparent

---

12 For a review of suggested phylogenies and definitions for three species of *Australopithecus* see Tobias 1968b; also Robinson 1968; Campbell 1972; Pilbeam 1972, inter alia.
13 This was not agreed to by Louis Leakey, whose hypothesis (see for example 1972) was taxonomically and phylogenetically clear whatever its merits: *H. habilis* was the paleospecies leading directly to *H. sapiens, H. erectus* being a contemporary side branch.

variation in evolutionary configuration shown, by the way, in good specimens. Outstanding is the now well-known ER 1470 which in some ways suggests an over-blown 'gracile' australopithecine with both a relatively large brain and a large face and teeth, but which is considered by Richard Leakey to be a very early specimen of *Homo*. Some three-quarters of a million years later there has appeared a range of specimens from (1) ER 1813, which Richard Leakey considers (1976b) probably to represent *Australopithecus africanus*, i.e. a standard gracile like the South African form, through (2) a skull which looks (to me) like a transitional mosaic of advanced australopithecine and robust *H. erectus* traits (ER 1805), although it is not the kind of intermediate that ER 1470 suggests, to (3) the fine skull that Leakey and Walker (1976) have frankly assigned to *Homo erectus* (see above). These last three forms are apparently too nearly contemporaneous to be arranged in a chronologically logical evolutionary sequence.

Most people have been cautious in assigning other recent cranial, mandibular, or dental finds. Richard Leakey and his associates have been chary of doing more than assigning them to probable *Homo* of indeterminate species. Others approach taxonomy with more confidence. For example Colin Groves, a practicing primatologist and paleontologist, believes (Groves and Mazak 1975) that the East African material can be segregated to show two nonsympatric but contemporaneous species of gracile *Homo*. He accepts *Homo habilis* as originally constituted present at Olduvai in beds I and II and also in the Omo valley. Contemporary upper Ileret, East Turkana, parts, mostly mandibular, he puts by partly metrical criteria in a new species, *Homo ergaster*. This includes certain jaws that to others had seemed more *Homo*-like than *H. habilis*, though placed in the latter species by Richard Leakey (according to Groves). Groves and Mazak also recognize *Homo africanus* of South Africa on Robinson's terms (i.e. *Australopithecus africanus*).

In this interpretation we have three species of *Homo* which, although of early date, are not *H. erectus*, but which otherwise cover everything in Africa that is not a robust australopithecine. But even *Homo habilis*, or other vague specimens of *Homo*, pose the question: with the appearance of a convincing (and well-preserved) specimen of *Homo erectus* in ER 3733, at 1.6 to 1.8 million years, what do these other specimens, of the same or later date, really mean? Are they in fact *Homo*?

There are also the much earlier specimens. Mary Leakey has found mandibles suggested to be *Homo* at Laetolil with a date of over 3 million years. The humeral fragment from Kanapoi, with a date of about 4.4. million, could not be distinguished from *Homo sapiens* morphologically or by multivariate analysis by Patterson and myself in 1967 (or by much more searching analysis by others since then). We suggested that it might represent *Australopithecus* because at that time allocation to *Homo* seemed preposterous, although it would be the correct one without the time element. (Of course, 'not different from' is not equal to 'the same as'). Finally, while the status of the already rich material from the Afar in Ethiopia dated at 3 million years is not yet clear (both a gracile and a robust australopithecine? [Johanson and Taieb 1976; Taieb et al. 1976]), it includes specimens assigned to *Homo*, especially two palates with teeth. These are said to resemble Sangiran 4 (*Pithecanthropus* IV) in many details, including precanine diastemata. From the descriptions and illustrations one may judge that if they had been found in Java they would have been referred at once to that group known as *Homo erectus* (*Pithecanthropus*) *modjokertensis* of the Puchangan deposits. (Here, as bearing on definition, we might raise the question whether these early forms may actually not be *Homo erectus*, but lower in grade; Corruccini [1974], analyzing calvarial shape [not maxillary features] by multivariate analysis, finds *Pithecanthropus* IV placed with australopithecines, not with Trinil and other *H. erectus*.)

The Problems of Asia
There can be little doubt that Africa was the original home of hominids. Many lines of evidence show the mutually close relationship of gorilla, chimpanzee, and *Homo sapiens*. But does this extend to all the transitions leading to the last-named, especially the emergence of *Homo erectus*? It has been widely assumed that the *Australopithecus*-to-*Homo* step was in Africa and Leakey and Walker (Leakey 1976a; Leakey and Walker 1976), although cautious on the last point, find no evidence for the existence of *Australopithecus* outside Africa, a finding that strongly bolsters the case. But *H. erectus* (or at least the Djetis zone hominid) was present very early in Java (and China?) — so early that some have found the admittedly rather isolated dates of about 1.9 million years difficult to accept (Butzer and Isaac 1975). (However, the first K/Ar dates of similar age for early Bed I at Olduvai were also greeted dyspeptically.) How should this be regarded?

A few present indications question a purely African line of emergence for *Homo*. One is the recent study of 'distances' among higher Old World primates measuring genetic divergence over time by the technique of DNA hybridization (Benveniste and Todaro 1976; Wong-Staal et al. 1976). One set of distances, based on general cellular DNA, corresponds entirely with other accepted schemes of monkey and hominoid groupings, including the particular closeness of the man-chimp-gorilla group. The other set is based on a special component of DNA, the gene for Type C endogenous virus, measured in distances from the baboon. These divergences are considered to be distorted by a selective environmental factor; release of Type C virus by baboons into the environment with possible infectious effects is supposedly defended against in other African primate species by maintenance of their own endogenous Type C virogenes in structure closer to that of the baboon. In the 'distances' found, Asian members in any branch are farther removed from baboons than African members (e.g. langur vs colobus), and man is equally distant with orang, gibbon, and siamang, while the African apes are much closer to baboons. This is not taken to mean that man is phylogenetically closer to the orang, but rather that the two reflect the early divergence of hominoids from all cercopithecids (including the baboon), as manifested in the cellular DNA, while the special Type C virogene divergence has been environmentally inhibited in the African apes. The conclusion is that the human lineage, like the orang's, has been isolated from the sub-Saharan African (baboon-contaminated) viral environment for some millions of years, and has reentered Africa too recently to have shown a genetic response. This of course can be positively applied only to living man, who is the subject of the tests, but necessarily implies that his later phylogeny is essentially non-African.

Secondly, Oxnard (e.g. 1975; Lisowski et al. 1976) and some fellow-conspirators are suggesting, from multivariate studies of postcranial anatomy, that the australopithecines were simply an experiment that failed. Oxnard concludes from a part-by-part analysis that function of these parts was not human or near-human, not placeable in a simple man-to-African ape scale. Rather, in a multivariate space, they are either uniquely different from both man and *Pan*, or veer somewhat in an orang direction. None of this is to deny their hominid affiliation; it is just that they cannot be read. Oxnard et al. assess the tali of *Homo habilis* and *Paranthropus* as 'totally non-human and possibly even non-bipedal.' They do not write off everything in East Africa. The 4.4-million-year Kanapoi humerus continues, in much more complex studies than the original, to be identified as *Homo*. But australopithecines, including *H. habilis*, are relegated to the status of 'pseudoman.' These are strange and unwelcome ideas, and some have been promptly disputed by other knowledgeable people (McHenry 1975; Rhoads and Trinkaus 1977). But Oxnard is an old hand at this business, having refined his material and methods over some time. One should take note of his views, clutching a saltcellar for protection.

Another aspect that questions a purely African line of emergence for *Homo* is simply the unknown. The present pattern of understanding is little better than the geography of discovery. East Africa is now overwhelmingly impressive in what it has provided. But we may remember where the first finds of earlier hominids were made — in Java and South Africa. The chances that either of these was the actual seat of emergence of *Homo erectus* (in spite of the principle that promising allopatric clades are apt to arise in peripheral isolates) are extremely small. Both are unambiguously marginal, and Pliocene deposits in Java are in fact primarily marine, which is scarcely a hominid environment.

East Africa, of course, is another matter. But southern Asia is a blank at the moment. China has been disappointing. As Clark Howell put it after a recent visit, why have no more Chou Kou Tiens been discovered? However, a fossil seam of the right date might change the picture catastrophically. In the realm of hypotheses, species or subspecies of *Homo*, developing in various Asian localities and migrating occasionally into Africa, might lead to the picture of finds as seen recently, with the corollary that any such immigrants would have been at least specifically distinct, and reproductively isolated, from australopithecines. (Sympatry of closely related animal taxa must usually derive from invasion or migration of an allopatrically evolved member.)

I do not put much stock in the above. I introduce it as a reminder that, as in the past, certain propositions may be widely subscribed to, which might conceivably prove mistaken. One such proposition is surely the African origin of *Homo*.[14] Another is the notion that hominids, homogenized by culture and adaptive flexibility, were not speciating at a time when elephants, equids, and suids were doing so freely. ('Speciating' would indeed be a strong word for hominids, but there was probably a strong tendency to subspeciate, even while such subspecies remained in the limits of what Simpson calls a genetical species, i.e. the normal biospecies.) Another notion might be the supposed early and gradual development of speech. In spite of some doubts that have been expressed about this, Lieberman and associates (e.g. Lieberman et al. 1972) have not had an appreciative audience for their work suggesting that the classic Neanderthals were distinctly limited in the ability to articulate as compared with modern man (see e.g. Burr 1976).

CONCLUSIONS

I have not intended to define *Homo erectus* afresh but only to suggest by my review that this should be done and what should be considered in doing so. Nor have I meant to assess the validity of *H. habilis* (even by putting it in quotes), or

---

14 This is what the extraordinarily percipient Darwin wrote in 1871: 'it is ... probable that Africa was formerly inhabited by extinct apes closely allied to the gorilla and chimpanzee; and as these two species are now man's nearest allies, it is somewhat more probable that our early progenitors lived on the African continent than elsewhere ... Nor should it be forgotten that those regions which are the most likely to afford remains connecting man with some extinct ape-like creature have not as yet been searched by geologists' (Darwin 1874).

other proposed species of *Homo*, of which at least six may be found in current writings of responsible scholars. With these, as well as several named species of australopithecines, we may be getting back toward the situation of a generation ago, not in irresponsible naming, but in staking out divergences of interpretation in fixed taxonomic terms. In any case, all of us are now aware, with centuries of discoveries still ahead, of the probabilities that any given present hypothesis is more likely to turn out wrong than right, which, however, is no reason for not hypothesizing.

To return to the main question, what is the meaning and usefulness of *Homo erectus* just now, as a taxon and concept? Of course it is meaningful and useful as long as it is not misused. Some problems come from attempts to classify, by naming, material which is too fragmentary to be so treated. Of course, some giving of taxonomic names to hominids is needed in order to organize the fossils as far as possible, but this is not an end in itself. Rather, in the present state of material, it should be used as a disciplinary device in evaluating interpretations.

For example, *Homo erectus*, unless one refuses to separate it from *Homo sapiens*, is obviously a perfectly good species, resting on the basic Chou Kou Tien collection and on its original type in the now expanding Java Trinil zone material. On present grounds it is represented in Africa in ER 3733, which cannot be excluded from it morphologically. But what about contemporary though fragmentary remains anywhere, which also cannot be excluded as far as they go? Parallel or diverging subspecies are one thing, if demonstrable, but *Homo erectus* is of no use as a wastebasket category; this is an objection to viewing the species as a grade, as it has been to the idea of a 'Neanderthal' stage.[15] Similarly, in East (and South?) Africa there are other early advanced or '*Homo*-like' finds, not including *H. habilis*, which can neither be put confidently in *H. erectus* nor used to define other species of *Homo*, such as might shed light on the early stages of the genus.

The arguments about *Homo habilis*, which is now seen in any case to be contemporary with *H. erectus*, also show the problem of drawing lines. Tobias, for example, who recognizes the species, sees evolution primarily in terms of grades (e.g. 1968b, 1973). *Australopithecus*, he states, gave rise to the 'time-successive species' *Homo habilis*, *H. erectus*, and *H. sapiens*. In his view, however, this is not a simple anagenetic transformation. There were constant cladogenetic trends, giving rise to 'much polytypy,' which would seem to correspond to other views of very broad variation at any level. But perhaps this

---

15 In any case, 'grades' have no taxonomic standing, however useful they may be in discussions of interpretation.

'reticulate evolution' (Tobias 1973) does not sufficiently acknowledge cladistic divergence of the kind recognized in the 'punctuated equilibria' pattern.

Simons, Pilbeam, and Ettel (1969), even before recent finds in East Africa. were among those who questioned the validity of *Homo habilis* as being based on material not clearly distinct from South African *Australopithecus*. They suggested procedures for the taxonomic placement of new hominid finds, briefly as follows: make sure a specimen is distinct from previously described contemporaneous species; if a good species, determine what *lineage* it belongs in; decide at what points in the lineage specific (or generic) distinctions are warranted. If the information on a new find is inadequate for these purposes, it should remain unnamed or tentatively assigned to an existing taxon. These are good rules, by which *Homo erectus* stands and *Homo habilis* wobbles. They imply lineages and not only grades. Thus if *habilis* stands in no lineal (ancestral) relation to *Homo erectus*, what is the criterion for putting it in *Homo* rather than in *Australopithecus*? Hominid taxonomy in general should look for the possibility of more than one lineage of the ancestral species or genus from which *H. erectus* survived. Similarly, this pattern should be looked for in the later evolution of *H. erectus*. If the Neanderthals should turn out to be an individual clade derived directly from the *erectus* stem and not through a *sapiens* stage that was also ancestral to living man, would it be correct to place them in *Homo sapiens* at all?

So *Homo erectus* is an important pivot in judging patterns of interpretation, against which early, contemporary, and late fossil hominids can be assessed. Its real nature needs to be further defined, and it should not slip into the idea of a classificatory convenience or be viewed entirely as a stage leading toward modern man. *Homo erectus* was a successful animal in its own right.

ADDENDUM

I have not attempted to rewrite this chapter to accommodate new materials and publications since the symposium was held and manuscripts were submitted. Some significant new finds have been announced or described, and the period has been especially productive in critical and analytical papers, above all in 1978 and 1979. This contribution was meant to be an essay of a general kind, for the purposes of the symposium, and the ideas expressed have not been greatly affected by later developments. A complementary article, addressed more to substantive matters, is in preparation for the *Yearbook of Physical Anthropology 1980*.

A few important developments may be mentioned as modifying matters in this paper. One is the placing of all the early Afar material in *Australopithecus afarensis*, a single species with wide size variation, by Johanson and White, thus eliminating any question of the presence of *Homo*. Clark Howell has published, in 1978, an extensive and detailed review of African hominids, with definitions for genera and species. Important later Middle Pleistocene remains were found in north China in 1976 and 1978, while further recoveries by Jacob in Java have extended the Sangiran and Ngandong samples, and promise to give better clues to the dating of the latter.

6

TEUKU JACOB

Solo Man and Peking Man

Cette étude décrit les caractéristiques et les variations morphologiques des pithécanthropes indonésiens, y compris *Pithecanthropus modjokertensis*, *P. soloensis*, et *P. erectus*, notamment leurs différences crâniennes, et comparent ceux-ci avec *P.* (=*Sinanthropus*) *pekinensis*. Les raisons de l'inclusion de l'homme de Solo dans les pithécanthropes y sont exposées et de nouvelles trouvailles provenant du même site y sont signalées. Il y a une ressemblance plus grande entre les types de Solo et de Pékin qu'entre ceux de Solo et de Java. D'importantes caractéristiques observées sur la base crânienne et le moulage endocrânien sont aussi présentées.

L'étude démontre également que le cannibalisme comme pratique courante, à des fins alimentaires ou rituelles, était peu probable chez les populations de l'homme de Solo et de l'homme de Pékin. Cette conclusion est tirée en se plaçant au point de vue démographique et en se basant sur l'étude anatomique des fractures de la base crânienne observées sur les ossements.

INDONESIAN PITHECANTHROPINES

Most of the early man fossils found in Indonesia are pithecanthropines. Most are skulls or their fragments (35 specimens), an endocast, and thighbones or their fragments (5 specimens), although the affinities of the latter are doubted (Day and Molleson 1973; von Koenigswald and Weidenreich 1939). In the cranial group we have 1 skull, 4 calvariae, 12 calottes, cranial fragments of at least 12 individuals, maxillary fragments of 2 individuals, mandibular fragments of 4 individuals, and isolated teeth (Jacob 1976a; Oakley et al. 1975). A complete list of the material is given in Table 6.1.

Pithecanthropine skulls are characterized by the underdevelopment of the cerebral cortex, the lower degree of basicranial deflection, and the robustness of

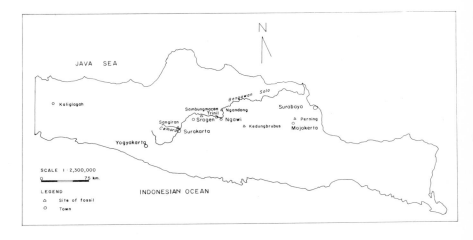

FIGURE 6.1
Paleoanthropological sites in Central and East Java. (Prepared by O. Soedarmadji.)

the masticatory apparatus relative to modern man. While the brain is larger in size and more developed than in *Australopithecus*, it is still relatively small, and the frontal, temporal, and parietal lobes are less developed than in *Homo sapiens*. On the external aspect of the skull these facts are reflected in the low position of the greatest breadth of the skull, the flatness or depression in the parasagittal regions, the low and receding forehead, the supraorbital torus and postorbital constriction, the small and low squama of the temporal bone, the small and low occipital plane, the angulated petrotympanic axis, and the absence of the foramen lacerum and petrooccipital fissure. On the internal aspect of the skull, the differences are reflected in the small cerebellar fossae, the relatively small anterior branch of the middle meningeal artery, and the vague impressions of the cerebral gyri and sulci (Jacob 1975a, 1976b).

In the endocast we observe the weakly curved, narrow frontal lobe and the well marked occipital pole. The height is small and the greatest breadth is located posteriorly and inferiorly. The interorbital portion protrudes farther forward, while the portion between the orbital surfaces is broader than in modern man, corresponding to the condition of the anterior cranial fossa observable in Ngandong 7.

The lower degree of flexure of the cranial base is revealed in the small height, the shape and position of the foramen magnum, the great nasion-basion distance, the long anterior cranial fossa, and the slight differences in depth of the cranial fossae. These facts can be noticed in the endocast as well.

The powerful chewing organs are reflected by the large cheek teeth, the heavy alveolar processes, the strong oblique lines of the lower jaw, the low position of the mental foramen, prognathism and the weak canine fossa, the pronounced supraorbital and occipital tori, the extensive nuchal area, the high position of the inion and the low position of the entinion, the prominent supramastoid crest, and the lateral position of the mandibular fossa (Jacob 1976a).

Robustness in *Pithecanthropus* is demonstrated by the well marked muscular crests on the skull and lower limb bones and by the thick vault bones (the mean of maximum thicknesses of the frontal, parietal, and occipital bones ranges between 11.4 and 13.2 mm). The last-mentioned trait is interesting because the thickness of the bones of the cranial base (except the tympanic plate) and parts of the temporal squama are the same as in modern man. Thickness of vault bones is also encountered in certain pathological modern skulls or in a few skulls of subrecent skeletons, but in the pithecanthropines all vault bones that have been discovered are consistently thick.

Pithecanthropus has a stature varying between 165 and 180 cm. The cranial capacity ranges from 750 to 1300 cm^3, which is, as stated by Dubois (1894), two-thirds that of modern man, with an average of 950 cm^3.

Three groups of pithecanthropines can be recognized in Indonesia. The first one is *Pithecanthropus modjokertensis* found in the Lower Pleistocene of Sangiran and Perning. The second, the largest group, is *P. erectus* found in the Middle Pleistocene of Sangiran, Trinil, and Kedungbrubus. The third is *P. soloensis* found in the Middle Pleistocene of Ngandong, Sambungmachan, and Sangiran (Jacob 1973, 1975b). Whether the differences are at the specific or subspecific level is at the moment difficult to judge because of the scarcity of the material for population studies, especially of *P. modjokertensis*. But it is unlikely that their grouping is due to age and sex differences; if this were so, then at Ngandong only male fossils were deposited, and in the Kabuh formation of Sangiran only one male fossil was preserved among many female ones. As for the possibility of age differences, we have, in fact, skulls with open and with completely obliterated vault sutures in each group.

P. modjokertensis and *P. soloensis* are more robust than *P. erectus*. They have a larger cranial capacity (1000-1300 cm^3), thicker vault bones, an occipital torus with a triangular prominence continuing as external occipital crest, sign of breakdown of the curved supraorbital torus into two halves, suprameatal spines, stronger supramastoid crests and suprameatal tegmen, and a different shape of the mandibular fossa. Both have a deep mandibular fossa with a higher posterior wall and the petrotympanic fissure running through the apex of the fossa. In *P. modjokertensis* (S 4) the articular tubercle is not well developed, and neither is the ectoglenoid process; the entoglenoid process is built up by the sphenoid

90 Homo erectus

TABLE 6.1
Pithecanthropine finds from Indonesia according to year of discovery*

Year of discovery	Site	Code	Stratigraphy†	Finds
19th century				
1890	Kedungbrubus	Kb 1	Kabuh Formation	Right mandibular corpus, juvenile
1891	Trinil	T 2	Kabuh	Calotte
1892	Trinil	T 3	Kabuh	Left femur
1900‡	Trinil	T 6	Kabuh	Right femoral fragment
1900‡	Trinil	T 7	Kabuh	Left femoral fragment
1900‡	Trinil	T 8	Kabuh	Right femoral fragment
First half 20th century				
1931	Ngandong§	Ng 1	Ngandong High Terrace	Calotte
1931	Ngandong	Ng 2	Ngandong	Frontal bone, juvenile
1931	Ngandong	Ng 3	Ngandong	Calotte
1931	Ngandong	Ng 4	Ngandong	Left parietal fragment
1932	Ngandong	Ng 5	Ngandong	Calotte
1932	Ngandong	Ng 6	Ngandong	Calotte
1932	Ngandong	Ng 7	Ngandong	Calvaria
1932	Ngandong	Ng 8	Ngandong	Right parietal fragment
1932	Ngandong	Ng 9	Ngandong	Right tibia
1933	Ngandong	Ng 10	Ngandong	Right tibia
1933	Ngandong	Ng 11	Ngandong	Parietal bones
1933	Ngandong	Ng 12	Ngandong	Calotte
1933	Ngandong	Ng 13	Ngandong	Calotte
1933	Ngandong	Ng 14	Ngandong	Calvaria
1936	Perning	P 1	Puchangan Formation (K/Ar, Pm)¶	Calvaria, juvenile
1936	Sangiran	S 1b	Puchangan	Right mandibular corpus
1937	Sangiran	S 2	Kabuh	Calotte
1938	Sangiran	S 3	Kabuh (Notopuro∥?)	Calotte
1938-39	Sangiran	S 4ab	Puchangan	Calotte and maxilla
1939	Sangiran	S 5	Puchangan	Right mandibular corpus
1937-41	Sangiran	S 7	Kabuh/Puchangan	Teeth
Second half 20th century				
1960	Sangiran	S 9	Puchangan	Right mandibular corpus

TABLE 6.1 continued

Year of discovery	Site	Code	Stratigraphy†	Finds
1963	Sangiran	S 10	Kabuh (K/Ar, Pm¶)	Calotte and left zygomatic bone
1963	Sangiran	S 11	?	Teeth
1965	Sangiran	S 12	Kabuh (K/Ar, Pm¶)	Calotte
1965	Sangiran	S 13ab	Kabuh	Parietooccipital and cranial fragments
1966	Sangiran	S 14	Kabuh	Basicranial fragments
1968-69	Sangiran	S 15ab	Kabuh	Left and right maxillary fragments
1969	Sangiran	S 16	?	Teeth
1969	Sangiran	S 17	Kabuh	Cranium
1969-70	Sangiran	S 18ab	Kabuh	Parietal and occipital fragments
1970	Sangiran	S 19	Kabuh	Occipital fragment
1971	Sangiran	S 20	Kabuh	Temporal and parietal fragments
1973	Sambungmachan	Sm 1	Kabuh (Pm¶)	Calotte
1973	Sangiran	S 21	Kabuh	Gonial region of mandible
1974	Sangiran	S 22	Puchangan	Mandibular corpus and maxillary fragment
1975	Sangiran	S 23	Kabuh	Endocranial cast
1976	Ngandong	Ng 15	Ngandong	Cranial fragments
1976	Ngandong	Ng 16	Ngandong	Calotte

* Meganthropine finds are not included.
† Puchangan: Lower Pleistocene; Kabuh, Notopuro, and Ngandong: Middle Pleistocene
‡ Dubois (1932)
§ All fossils from Ngandong were found in excavations.
¶ Pm: Paleomagnetic dating is being carried out.
∥ von Koenigswald and Ghosh (1973)

bone. The tympanic plate is almost vertical and has a strong tympanic crest. On the other hand, in *P. soloensis* the articular tubercle is larger and the posterior wall extends farther down, and both the ectoglenoid and entoglenoid processes are well developed. The entoglenoid process is formed by the sphenoid and temporal bones.

Other continuous and discontinuous traits differentiate *P. modjokertensis* and *P. solensis*. The first mentioned has a slightly smaller cranial capacity, a large paramastoid eminence, elliptical, thin-walled, and vertically oriented external ear holes, and a relatively lower petrous pyramid, especially in the middle cranial fossa. In *P. soloensis* the foramen ovale is situated in a fossa with an accessory oval foramen that is sometimes confluent. Endocranially there is a groove from the accessory foramen to the foramen rotundum and the trigeminal impression; I think that it serves as a canal for the accessory meningeal artery instead of for the motor root of the trigeminal nerve (Jacob 1967).

P. erectus is gracile; he has a smaller cranial capacity (750–1000 cm^3), a continuous and straight supraorbital torus with pronounced postorbital constriction, a rounded occipital torus, and no external occipital crest. The mastoid processes are small and no suprameatal spine is present. The zygomatic bone is also small and gracile. In *P. erectus* the mandibular fossa is not deep and the articular tubercle is not distinct. The posterior wall is not high and the ectoglenoid process is poorly developed. The petrotympanic fissure is on the posterior wall. Furthermore, the entoglenoid process is formed by the sphenoid bone.

P. modjokertensis lived around 1.9 million years ago, and *P. erectus* from 900,000 to probably 600,000 years ago. *P. soloensis* lived at the same time as *P. erectus* but survived up to probably 200,000 years ago (Jacob 1975b). It seems that both *P. erectus* and *P. soloensis* evolved from *P. modjokertensis*, and it might be that it was *P. erectus* that further evolved anagenetically towards *H. sapiens*.

SOLO MAN, PEKING MAN, AND JAVA MAN

Of the three groups mentioned above the one most frequently compared with *Pithecanthropus* (*Sinanthropus*) *pekinensis* is *P. erectus*. This is not surprising because when the first fossils of Peking Man were discovered, remains of *P. soloensis* and *P. modjokertensis* had not yet been found. It was Weidenreich (1937a, 1937b) and von Koenigswald and Weidenreich (1939) who made extensive comparisons between *P. pekinensis* and *P. erectus*. *P. modjokertensis* is represented only by one adult specimen and one juvenile, and hence is unsuitable for comparative study. *P. soloensis* has not often been taken into consideration because most people believed that it was an equatorial Neanderthal Man or even *H. sapiens sapiens* owing to its supposedly late date.

Weidenreich (1943, 1951) subsequently compared *P. soloensis* with *P. pekinensis* only because of the similarities of the latter to *P. erectus* and its hypothesized evolution into *H. sapiens* of the region. Actually Weidenreich thought that *P. pekinensis* evolved into Mongoloids through the Chou Kou Tien Upper Cave people, whereas *P. erectus* and *P. soloensis* evolved into the Australoids.

Dubois (1936) did not believe that any other human fossils found after the Trinil discovery were *P. erectus*. He classified Sangiran 2 as Solo Man and related

P. pekinensis in the same way. This fact is very interesting because he mentioned it when it was widely believed that *P. soloensis* was Neanderthal. Disregarding his personal belief, in retrospect it is not surprising at all considering the pithecanthropine features exhibited by *P. soloensis*; it is more of a surprise why he did not include the Trinil skull cap in Solo Man as well. Hrdlička (Weidenreich 1943) also thought that the *Sinanthropus* Skull E was Neanderthal, just as Oppenoorth (1932, 1933) thought this of his Solo Man.

Ariëns Kappers (1929, 1936) noted the similarities of endocranial casts among *P. erectus*, *P. pekinensis*, and *P. soloensis*. Weidenreich elaborated the pithecanthropine nature of Solo Man. He noted in the latter, 55 of 58 features that resemble *P. erectus* more than Neanderthal Man. Others have recognized those similarities too, such as Coon (1962), Kurth (1965), and Mayr (1963).

In a complete study of the original fossils of Solo Man in 1967 I indicated their pithecanthropine traits as follows:

1 the cranial capacity is in the range of *P. pekinensis*;
2 the cranial contour in occipital, lateral, vertical, and frontal views are pithecanthropine;
3 the greatest breadth of the skull is located near the base;
4 the high position of the external occipital protuberance and the low position of the internal occipital protuberance;
5 the large size of the posterior branch of the middle meningeal artery and its low ramification from the common trunk that passes through a bony canal in all intact specimens;
6 features on the cranial base such as the petrotympanic angulation, the absence of the foramen lacerum and petrooccipital and sphenopetrous fissures, etc;
7 the shape of the individual cranial bones; and
8 the pattern of pneumatization.

Other authors have also pointed out other similarities between Solo Man and the pithecanthropines. Olivier and Tissier (1975) found that *P. erectus*, *P. pekinensis*, and Solo Man (Ngandong) show similar relationships between cranial capacity and the product of cranial dimensions.

In the last decade other specimens of *P. soloensis* have been discovered. The first was from Sangiran in 1969 (S17). It is a cranium which gives the first evidence of the face of Javanese pithecanthropines. It is very robust, has very thick calvarial bones and cheekbone (thickness 16 mm, height 41 mm), and strong muscular crests (Jacob 1973). The facial breadth is ca 158 mm and the moderately broad pyriform aperture displays no anterior nasal crest. The specimen came from the Middle Pleistocene Kabuh formation.

The second specimen is a calotte found at Sambungmachan on the bank of the Solo River in February 1973. It was discovered in a sandstone layer in the Middle Pleistocene beds, 4.74 m below the surface. It is not as robust as S17, but both skulls are undoubtedly *P. soloensis* because of the following characteristics (Jacob 1973, 1975a):

1. the cranial capacity of ca 1000 cm^3;
2. the cranial contour in various views;
3. various cranial dimensions;
4. the shape of the occipital torus;
5. the large mastoid processes;
6. the morphology of the external auditory meatus; and
7. the morphology of the mandibular fossa.

The third specimen was found at Ngandong not far from the original Ngandong site excavated by Oppenoorth in 1931-33. It consists of the right supraorbital portion of the frontal bone and fragments of the sphenoid and right temporal bones. The specimen was discovered on August 13, 1976, in an excavation of the high terrace of Ngandong (24.5 m above the Solo River, 54 m above sea level). The fossil lay in a gravel bed, and in the same square (of 2 x 2 m) 160 fragments of animal bones were recovered. Additional fossils were found in another excavation on December 30, 1976, and comprise the frontal, the sphenoid, the occipital, and the left temporal and parietal bones. They lay merely ca 2 m away from and 10 cm below the earlier find. In the same square 107 fragments of mammalian fossils were recovered. The animals were mostly *Bubalus*, *Bibos*, and deer. I have the impression that the two hominid specimens could belong to one individual, labelled Ng 15.

Ng 15 demonstrates a supraorbital torus that is clearly *P. soloensis*, and the same is true with the occipital torus. The fossilization is different from the earlier Ngandong skulls; there is marked infiltration of sandstone into the diploe.

Thus, we now have in total remains at least 17 individuals of Solo Man. As far as identification is possible, they represent 12 adults and one juvenile, five males and eight females.

If the antiquity of *P. lantianensis* is 700,000 years and that of *P. pekinensis* 400,000 years, then they can be chronologically compared with respectively *P. erectus* (with a radiometric date of 830,000 years) and *P. soloensis* of Ngandong which according to my estimate is at least 200,000 years old. Usually *P. lantianensis* has been compared with *P. modjokertensis* (with a radiometric date of 1.9 million years), and, as mentioned previously, *P. pekinensis* with *P. erectus*. It would be interesting, therefore, to look more closely at the similarities and differences among *P. soloensis*, *P. pekinensis*, and *P. erectus*.

TABLE 6.2
Cranial measurements in Solo, Peking, and Java men (in mm)

Measurement	Solo*			Peking†			Java‡		
	n	Mean	S.D.	n	Mean	S.D.	n	Mean	S.D.
Cranial length	6	202.1	9.7	5	193.7	3.7	4	180.1	3.4
Cranial breadth	6	148.0	4.2	4	140.3	2.1	4	137.8	6.2
Baison-bregma height	3	126.3	4.0	1	115.0	–	2	105.0	0.0
Auricular height	6	109.2	5.2	4	99.5	4.6	3	94.3	5.6
Cranial index	6	73.0	4.0	4	72.2	0.5	3	72.8	2.9
Length–basion-bregma height index	3	62.0	2.5	1	59.6	–	2	58.5	1.5
Length–auricular height index	7	54.9	2.9	5	50.9	1.4	3	52.0	2.4
Breadth–basion-bregma height index	3	87.1	1.7	1	75.6	–	2	79.3	2.1
Breadth–auricular height index	7	74.5	3.2	4	70.3	2.4	3	71.4	2.5
Cranial capacity	3	1210.0	81.9	5	1043.0	100.6	3	883.3	82.5
Minimum frontal breadth	7	106.6	3.3	5	85.9	3.5	3	82.0	2.4
Transverse frontoparietal index	6	72.6	1.0	4	61.7	1.8	3	63.6	2.9
Bistephanic breadth	8	108.8	6.2	3	87.3	11.2	3	84.0	10.6
Biasteriac breadth	8	138.1	10.0	5	111.8	4.8	3	112.7	14.7
Transverse parietooccipital index	6	95.9	1.9	4	81.7	2.9	3	87.2	10.1
Bimastoid breadth	2	110.5	3.2	2	104.5	1.5	2	108.0	6.0
Nasion-bregma chord	6	116.4	1.9	5	109.8	5.0	3	91.7	4.5
Nasion-bregma arc	6	128.5	3.5	5	122.6	4.5	2	95.0	5.0
Frontal sagittal curvature index	6	90.6	1.8	5	89.9	1.7	2	97.9	0.2
Bregma-lambda chord	8	101.8	8.1	5	96.2	7.7	4	89.8	1.8
Bregma-lambda arc	8	106.0	7.6	5	103.9	7.8	4	93.5	1.5
Parietal sagittal curvature index	7	94.9	0.8	5	94.1	0.9	4	91.9	2.7
Lambda-opisthion chord	6	85.2	2.5	3	84.0	2.8	4	99.5	3.4
Lambda-opisthion arc	6	116.7	6.0	3	114.0	5.7	3	105.7	5.2
Occipital sagittal curvature index	6	72.0	2.3	3	73.8	0.7	3	74.7	0.7
Biporion chord	6	125.5	3.8	4	123.7	2.9	3	126.3	8.4
Biporion arc	4	302.8	8.5	4	286.8	13.5	3	263.3	5.0
Transverse curvature index	4	41.0	0.7	4	51.1	2.6	3	47.5	4.0

* Jacob (1967, 1976a); Weidenreich (1951)
† Weidenreich (1943)
‡ Jacob (1973, 1975a)

The Solo, Peking, and Java men are craniometrically compared in Table 6.2. Traits shared by all three species have been discussed at the beginning of this paper. It can be added that pithecanthropines have a strong frontal crest but no crista galli or foramen cecum. The crest separates the left and right frontal lobes. No sagittal groove is discernible in this area. The Sylvian crest is very well developed, but the petrous pyramid is low, relative to the cranial fossae. On the external aspect the dolichocranic skull discloses a similar occipital sagittal index, cranial breadth, and nasal breadth. Other similarities are revealed in the coincidence of the opisthocranion and inion, the absence of the styloid process of the temporal bone, the wide squamous suture, and the wide supraorbital notch.

Traits shared by *P. pekinensis* and *P. erectus* are as follows. Both display similarities in the frontoparietal and parietooccipital indices, the length-height and breadth-height indices, and the total sagittal curvature index. Several transverse measurements are also similar, such as the cranial breadth, minimum frontal breadth, bistephanic breadth, and biasterianic breadth. The occipital bone shows resemblances in the shape of the occipital torus, the occipital angulation, and the ratio between the occipital and nuchal planes. Furthermore, the size of the mastoid process and the zygomatic bone are the same in both.

Traits shared by *P. soloensis* and *P. erectus* are the bimastoid and biporion breadth, the small occipitomastoid crests, the presence of the vaginal process of the temporal bone, and the wide sigmoid sulcus. Both the orbital and occipital supratoral grooves are not as distinct as in *P. pekinensis*.

Traits shared by *P. soloensis* and *P. pekinensis* are shown in the cranial capacity, and the cranial length, the frontal bone (receding, although the nasion-bregma distance is greater in *P. soloensis*), the parietal bone (the acute parietal notch, the parietal sagittal index), the occipital bone (the height of the occipital plane), and the suprameatal spine. The supraorbital torus is curved above the orbits, and the orbital index is similar in both species, although we do not know the condition in *P. erectus*. We can also add that the superior orbital fissure is short in both and that the axis of the orbital cavity is long. A strong orbital crest is present on the sphenoid bone. The postorbital constriction is not as pronounced as in *P. erectus*.

There are some dissimilarities among the three species. Although the general shape of the skull is alike, the skull of *P. soloensis* is broader and that of *P. erectus* is lower than in *P. pekinensis*; the breadth-height index, therefore, varies. The supraorbital torus shows supraglabellar disintegration in *P. soloensis*, but in *P. erectus* it is continuous and straight from side to side. The lower margin of the pyriform aperture is also interesting. While there is a prenasal fossa in *P. soloensis*, an anterior nasal crest is present in the other two, but no anterior nasal spine is discernible in *P. pekinensis*. The details of the mandibular fossa are also dissimilar. In *P. pekinensis* it is deep, bounded by a well developed articular tubercle and an oblique posterior wall which extends downwards to the same level as the tubercle. The entoglenoid process is formed by the sphenoid bone, and the ectoglenoid process is also well developed. The conditions in *P. soloensis* and *P. erectus* have been described earlier.

It is surprising that the resemblance between *P. soloensis* and *P. erectus* is less than either between *P. soloensis* and *P. pekinensis* or between *P. pekinensis* and

P. erectus, although they were sympatric with some overlap in time. It would be interesting to compare *P. lantianensis* with *P. erectus* and *P. modjokertensis*. We need more radiometric dates of more sites to understand the evolution and migration of pithecanthropines.

A note should be added to what Weidenreich (1951) reported as the postjugular fossa and the large hypophyseal fossa in Solo Man. In my study (Jacob 1967) I observed the 'postjugular fossa' in Ng 6, Ng 13, and Ng 14, and because of its irregular shape, varied location, the thinness and diploetic appearance of its floor, its continuity with adjacent broken areas, its inconstant presence, and the presence of the real jugular fossa, I am convinced that the structure is artificial.

The large-sized hypophyseal fossa is also artificial as confirmed by examination with a throat diagnostic set. In Ng 7 the clivus, the posterior and anterior clinoid processes, the sellar tubercle, and the sphenoidal sinus are broken. In Ng 14 the dorsum sellae is also broken, as are the sellar tubercle, the chiasmatic groove, and the sphenoidal sinus. Therefore, the dimensions of the hypophyseal fossa in Solo Man are impossible to assess.

CANNIBALISTIC BEHAVIOR

Another characteristic that associates Solo Man and Peking Man is their supposedly cannibalistic behavior. The reasons for the claim are as follows (Jacob 1972):

1 Most fossils found are calottes or skull fragments.
2 The cranial base and the face are rarely found, and if found, they are usually damaged.
3 There are signs of physical trauma on several skull vaults.
4 Many mandibles are found and are mostly broken.
5 The limb bones are frequently broken at their ends or transversely, supposedly due to attempts to get the bone marrow.
6 There is ethnographic evidence that until recently cannibalism has been practiced in or near the general area where Indonesian pithecanthropines used to live, and the present population is assumed to have inherited some pithecanthropine genes.
7 *Pithecanthropus*, with his huge eyebrow ridges in front of his low skull vault and his massive protruding jaws, must have looked very brutish, and consequently, it is not unlikely that he fought his own people and ate them too.
8 Technology and morality are supposed to go together; if the former is high, the latter should be high too, and the contrary is also true.

9 Wars and violence have been occurring endlessly among modern men, and no nation can claim innocence of aggression and cruelty. This trait must therefore have been inherited biologically from our early ancestors through the australopithecines and pithecanthropines.

It seems that the hypothesis of prehistoric cannibalism has never been worked out in detail. We have not tried to answer the following questions. Were all early men cannibals, or were all pithecanthropines cannibals? Whom did they eat — only members of other groups or also members of their own group? Did they eat only adults, only children, or both? Did they eat hominid meat regularly or just occasionally; or asked differently, did they eat it as their daily meal or merely for ritual purposes? Was hominid meat their only source of animal protein or just a supplement in their daily diet? Did they roast the meat or eat it raw, and did they store it? These matters might be distasteful, but the problem of cannibalism in early man has been with us for quite a while, so that it is proper if we consider it in a detached manner.

Suppose we have a pithecanthropine population of 50 individuals, one-half of which is between 15 and 30 years of age, one-third below 15 years, and one-sixth above 30 years. The birth rate is 40 per mille and the death rate 25 per mille. Stature in adults ranges from 165 to 180 cm and the average adult body weight is 80 kg (Jacob 1975b). As a practitioner of cannibalism each individual would need 80 g of protein per day, 40 g of which should consist of animal protein if they follow the American standard. If one-half of this amount is supplied by hominid meat, which is of high biological value, each individual would get 20 g per day of hominid protein.

On the other hand, as a victim of cannibalism one individual would contribute 7200 g of protein, calculated after Garn and Block (1970). If 90 percent of the meat is digestible, then each victim yields 6500 g of protein. If the meat could be stored for a week before its odor becomes too offensive for early man, each man would provide 19 percent of protein for the group for a week. Let us suppose that they eat only adults, simply for economic reasons, since children would give less meat and are better preserved for later needs. There are in our pithecanthropine group 41 individuals above 15 years, and there will be only two births (we have assumed that *Pithecanthropus* has a shorter gestation period) and one death each year, so that the annual natural increase is about one individual. In ten weeks only 31 adults are left, a 25 percent decrease of meat source and meat eaters alike. In other words, there are less people to hunt and to hunt with.

Three months later only 21 adults are left and four months later only 11. In nine months, or eighteen months since the beginning of cannibalism, only one adult remains. Consequently, all the adults will be eaten in 1.5 years, or if all the 40 g of animal protein requirement is supplied by hominid meat, all the adults will be totally consumed in nine months.

In 1.5 years there will be only three births and two normal deaths, and the natural increase is therefore only one. If they do not eat members of their own group, but only members of neighboring groups who are also cannibals, they will finish eating adults from both groups, reciprocally, during the same time period. Even if the other group is not cannibalistic and the first group eats also members of their own group, all the adults will be totally consumed in three years.

Thus, by practicing cannibalism man has some short-term advantages, such as the availability of quality protein which provides all the essential amino acids, the possibility of consuming a large brain as a special paleolithic delicacy, the hunting grounds are more familiar to the hunters, and in the case of eating infants of your own group, the daily effort involves only food gathering, which falls on the shoulders of woman.

On the other hand, the source of this kind of protein is limited and the behavior of the hunter is known to the hunted, while in the surroundings an abundant meat supply is available as shown by many fossil remains of mammals with high meat:bone ratio. In the case of eating members of your own group, terror and fear will be created in the group that will affect its cohesiveness. For the sake of self-interest, cannibalism would be stopped as soon as the idea is implemented. Otherwise, instability and splitting of the group will result and it will vanish in no time.

Hunting your fellow hominines of a different group would be more difficult than hunting herbivores, and one needs to invent regularly new and more clever tricks because the other group is hunting you at the same time. This would be an unusual case of energy flow in an ecosystem. Hunting conspecific members of one's own community is nothing more than a perpetual warfare in which the victims are consumed, and is very costly in terms of hominid survival, considering that at the same time other carnivores are also after the prey. The instinct of homeostasis would soon act to preserve the group. Existence is impossible if the consumer and his food are the same.

There may have been occasional cannibalism; it is just like occasional killing. Whether one kills somebody to eat him does not affect the existence of the species if it takes place occasionally or does not result in negative population growth. Thus, enemies might be killed and eaten to cool off hatred or perennial vendettas or as social intimidation display. Ritual cannibalism might be practiced in which case a chosen victim is eaten to procure his spirit or to maintain harmonious balance with the supernatural. Dead relatives might be ritually eaten to inherit their power, but this does not affect the existing death rate. This kind of cannibalism, however, appears to have developed later in cultural evolution, perhaps after the beginning of the Mesolithic or Neolithic and the development of articulate speech, abstract thinking, and tradition in the disposal of the dead (Jacob 1967, 1972).

There is also a possibility of emergency cannibalism, for example, in time of starvation. The group whose members practice this, as we all know, is usually not regarded as cannibalistic.

As far as the evidence goes we have nothing to prove occasional cannibalism among prehistoric man. The base of the skull and the face with their irregularly shaped bones which are very thin in certain spots are easily broken into tiny pieces. In *Pithecanthropus* the bones of the skull vault are thick (except for some parts of the temporal squama), but those of the skull base are thin (except for the tympanic plate) and very vulnerable to postmortem physical trauma. The base, in addition, has numerous holes and fissures. Furthermore, it is very unusual to find complete and intact fossilized vomer, ethmoid, lacrimal, sphenoid, temporal, and occipital bones.

In secondarily deposited bone beds, such as those of Java, skulls have a good chance of being preserved. Owing to their round shape they can better survive the rough transportation than the long or irregularly shaped bones. If fragmented, vault bones can be more easily identified than, for example, broken short bones, ribs, or pelvis. During the transport or after being deposited, soil and water penetrate the skulls and their pressure is directed mainly to the loosely attached bones of the cranial base (Jacob 1972). Continual rolling would result in the absence of the face (Boaz and Behrensmeyer 1976). Carnivores as a rule do not eat the head of their victims and this is left intact.

Accidents might easily happen during a paleolithic hunting trip that affect the skull, particularly the vault. But it should be noted that trauma to the vault may result in fracture of the base. Mandibles naturally break at the points of angulation such as the symphysis menti and the junction between the corpus and the ascending ramus.

The damage made by men on the skull base for ritual purposes, such as observed in skull collections of head hunting tribes, is structured and demonstrates a pattern, while the cranial bases of *Pithecanthropus* have unstructured damage; the fractures follow the natural lines of least resistance. For extracting the brain from the neurocranium the best and easiest approach is through the sides of the skull. This is confirmed by artificial holes found in the parietotemporal areas of the skulls in the collection of headhunters. Of course, in already cleaned skulls, devoid of soft parts, the best approach to the cranial cavity is the base, but this approach is clumsy if applied on heads that still have soft parts, particularly those with thick necks, and if pursued by primitive paleolithic tools (Jacob 1967).

FIGURE 6.2
Occipital views of *Pithecanthropus pekinensis* skull (O-CH 14, cast) and *P. soloensis* skull (Sm 1).

Taking into consideration all of the above, cannibalism in early man should be seriously doubted.

CLOSING REMARKS

The present condition of the former Sundaland, which because of its size could be called the Sunda subcontinent, discourages many students of paleoanthropology from regarding it as the place of original evolution of the pithecanthropines and favors the vast land of China, although available radiometric dates suggest otherwise. We owe our first knowledge of *P. pekinensis* to the foresight and anatomical eyes of the great Davidson Black. Future finds of pithecanthropine remains in China and Indonesia and the region in between, as well as more radiometric dates from many Pleistocene sites, will solve the problems of affinities among the pithecanthropines of Asia. To Davidson Black we owe the nomen *pekinensis*, and we hope that the future will throw more light on the evolutionary relationship between Peking Man and Solo Man.

102 Homo erectus

FIGURE 6.3
Superior views of *Pithecanthropus erectus* skull (S 10) and *P. soloensis* skull (Sm 1).

FIGURE 6.4
Occipital bone of Sm 1, showing the thinness of the bone at the cerebellar fossa and its thickness at the internal occipital protuberance.

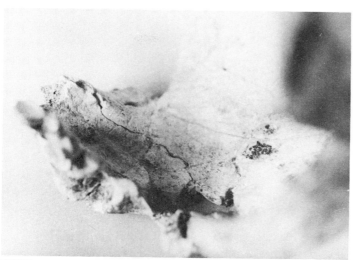

FIGURE 6.5
The right Sylvian crest of S 10 traversed by the meningeal canal for the middle meningeal artery.

FIGURE 6.6
Ng 15 skull found in situ in an excavation of the Ngandong terrace on August 13, 1976.

ACKNOWLEDGMENTS

Grants from the Department of Education and Culture, Jakarta, and the Wenner-Gren Foundation for Anthropological Research, New York, are greatly appreciated. I also appreciate the discussions with Dr I. Glover, Institute of Archaeology, London, and staff members of the Gadjah Mada University Department of Physical Anthropology, but the responsibility for the content of the paper is, of course, my own. The map was prepared by Mr O. Soedarmadji, formerly of the Department of Geology, Gadjah Mada University School of Engineering.

7

ANDOR THOMA

The position of the Vértesszöllös find in relation to *Homo erectus*

La première partie du présent article résume les caractéristiques morphologiques essentielles des Hommes mindéliens de Vértesszöllös (Hongrie). La seconde partie analyse la position taxinomo-phylétique des fossiles en question. Dans le système morphologique habituel les Hommes de Vértesszöllös se situent dans la zone intermédiaire entre *Homo erectus* et *Homo sapiens*. En appliquant les principes de la systématique phylogénétique de Hennig (1966) les Hommes de Vértesszöllös doivent être rangés dans à espèce *H. sapiens*.

In 1965, the Mindel period Vértesszöllös occupation site of the pebble/chopper tool culture in Hungary (Kretzoi and Vértes 1965) yielded remains of two hominid individuals, Vértesszöllös I and II. In this paper I elaborate on the substance of the remains, their morphological features in comparison with other fossil and modern hominid material, and on their phylogenetic and taxonomic positions.

MORPHOLOGY

Vértesszöllös I (Thoma 1967) is a child mainly represented by the crown of a lower left deciduous canine and the fragmentary crown of a lower left second deciduous molar. The crown of the canine is elongated mesiodistally as can be seen in comparison with data on modern man (Table 7.1). It has a distolingual cingulum. The lingual face is framed by a distinct enamel rim, and an enamel ridge runs from the base to the tip (Figure 7.1). All these features are Sinanthropoid characters.

The molar is straight-walled and mesiodistally elongated. It bears a classic *Dryopithecus* cusp pattern and a strongly developed *fovea posterior* but no

TABLE 7.1
Metrical comparison of the Vértesszöllös I deciduous canine with a recent series from the collection of the Hungarian Museum of Natural History, Budapest

		Length (m-d)	Width (b-l)	Robustness	Index
Vértesszöllös I		6.7	5.4	36.18	80.59
Recent	M	5.85	5.34	31.21	91.35
($n = 22$)	s	0.295	0.236	2.45	4.61
Difference/s		2.881	0.254	2.029	2.334
Probability	(%)	$1 > P > 0.1$	$90 > P > 80$	$10 > P > 5$	$5 > P > 2$

accessory cusps. Like the canine, the molar shows, if not complete identity, substantial structural conformities with the homologous teeth of *Sinanthropus* as reported by Weidenreich (1937a). It differs from the teeth of other hominids that were known at the time of my original study (Thoma 1967).

I adhered to this comparative morphological conclusion until 1975, and the scantiness of the material did not allow phylogenetic and taxonomic statements. In 1975, deciduous teeth of the Vértesszöllös type were found in a cave at La Chaise, France. The remains, that of a child from the Riss period, included parietal bones which displayed the bomb-shaped *Norma occipitalis* of Neanderthal skulls. I inspected these unpublished remains through the kind permission of their discoverer, André Debénath.

Vértesszöllös II (Thoma 1966, 1969) consists of the occipital squama (*squama occipitalis*) of an adult, probably male. The exocranial aspect, illustrated in Figure 7.2, shows a very high and strong occipital torus (*torus occipitalis transversus*), a distinct occipitomastoid crest (*crista occipitomastoidea*), and a small portion of the thickened rim of the foramen magnum (*limbus occipitalis*) which facilitates reconstruction of opisthion. As in the Swanscombe occipital, Waldeyer's crest is strongly developed bilaterally. The endocranial aspect (Figure 7.3) clearly shows the metasterionic angles and the smallness of the cerebellar fossae (*fossae cerebellares*) in relation to the cerebral fossae (*fossae cerebrales*). The cruciform eminence (*eminentia cruciformis*) and the sinus grooves are low and smooth. The groove of the transverse sinus (*sulcus sinus transversi*) passes directly to the temporal bone, while the endinion is situated deeply below the inion, indicating the lack of evolutionary rotation of the bone. The lateral aspect (Figure 7.4) shows a heavy backward protrusion of the occipital torus and a considerable thickness of the bone. The positions of lambda and opisthion were reconstructed by me and independently by Phillip Tobias who kindly allowed me to publish his results.

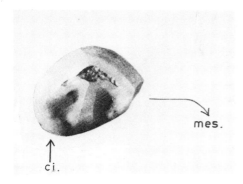

FIGURE 7.1
Vértesszöllös I: the left lower deciduous canine in incisolingual view; ci. = cingulum, mes. = mesial direction. Three-quarter size.

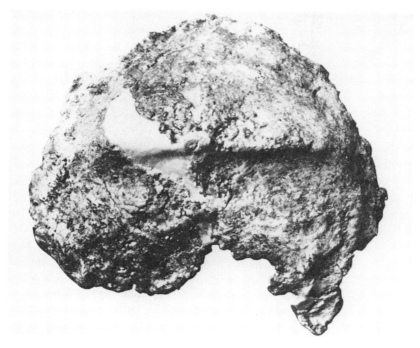

FIGURE 7.2
Vértesszöllös II: exocranial view. Three-quarter size.

108 *Homo erectus*

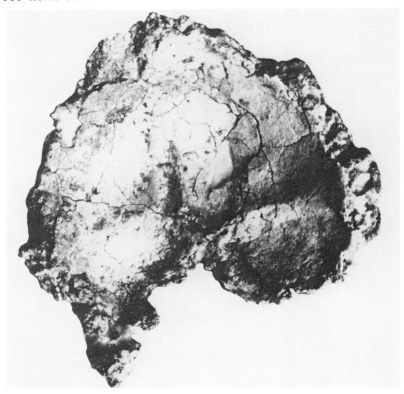

FIGURE 7.3
Vértesszöllös II: endocranial view. Three-quarter size.

As can be seen in Figure 7.5, Tobias' lambda position is identical with mine (l), but his opisthion (o_2) is 3 mm closer to inion. For topographical reasons, I feel that Tobias' reconstruction is better than mine (o_1). The sagittal section of the occipital bone further shows its wedge-like profile.

All the described characters of Vértesszöllös II are expressively archaic. The occipital plane (*planum occipitale*) is, however, high and fairly curved in the sagittal direction as in modern skulls. The shape of the occipital torus is unique. If the bony mass were reduced by a supratoral fossa (*fossa supratoralis*), a not illogical process for later evolution, it would be similar to the Swanscombe and La Quina tori.

I took measurements of the Vértesszöllös II occipital in 1965. Afterwards, the bone was badly damaged and deformed during preparation of a second series of casts. I estimated the cranial capacity of the individual by means of a regression

FIGURE 7.4
Vértesszöllös II: right lateral view. Three-quarter size.

equation found between the cranial capacity and lambda-opisthion chord (x) in a sample of eleven archaic fossil skulls (*Pithecanthropus* II and IV, *Sinanthropus* III, XI, and XII, Ngandong I, V, IX, and XI, and Swanscombe):

$$Y = 26.62x - 1199.4 \text{ cm}^3 \quad (s_Y = 60.9).$$

The criteria for selection of this sample were simply morphological; i.e. large biasterionic breadth, thick cranial walls, and lack of occipital 'chignon.' In the sample the correlation between cranial capacity and lambda-opisthion chord is $r = 0.929$.

All measurements, including cranial capacity, are reproduced in Table 7.2. Those starting from lambda or opisthion are reconstructed, while the remainder are of the intact bone.

Table 7.3 lists Penrose (1954) shape distances between each of nine fossil skulls and the medieval Zalavár series from Hungary (n = 109 males; Thoma

110 Homo erectus

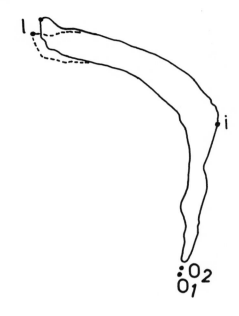

FIGURE 7.5
Sagittal section of the Vértesszöllös occipital bone. Three-quarter size. l = lambda, i = inion, o_1 = author's opisthion reconstruction, o_2 = Tobias' opisthion reconstruction.

1969) which represents modern man. In terms of numerical magnitude (the Penrose method gives no information about direction) three groups can easily be distinguished. The two Upper Paleolithic crania, the Oberkassel female and the Oberkassel male, are very close to the modern man sample, while the extremely archaic or specialized skulls, including *Sinanthropus* III, La Chapelle, and *Pithecanthropus* IV, are considerably distant. Vértesszöllös II is intermediate together with Skhul V, Swanscombe, and La Ferrassie I.

The general morphological characterization of Vértesszöllös II is, therefore, the following: a mosaic of modern and predominantly archaic features, an 'intermediate' metrical configuration, and a large cranial capacity (see Table 7.2) for the Middle Pleistocene.

TABLE 7.2
Metrical characters of the Vértesszöllös II occipital bone

Martin number	Character	$(l, -o_2)$
12	Biasterionic breadth	126.5
28	Lambda-opisthion arc	132
28(1)	Lambda-inion arc	79
30(3)	Lambda-asterion chord (left)	86
30(3)	Lambda-asterion chord (right)	91
31	Lambda-opisthion chord	100
31(1)	Lambda-inion chord	73
31(2)	Inion-opisthion chord	53
33(4)	Lambda-inion-opisthion angle	105°
38	Cranial capacity estimate	± 1300 cm³
–	Max. height of *torus occipitalis*	38
–	Inion-endinion distance	25
31:12	Height-breadth index	79.05
31(2):31(1)	Nuchal-occipital chord index	72.60
31:28	Sagittal curvature index	75.76
31(1):28(1)	Lambda-inion chord/arc index	92.41

Twiesselman number	Thickness	
6	5 mm below lambda	>10
8	*Fossa cerebralis sinistra*	± 10
8	*Fossa cerebralis dextra*	10
9	*Fossa cerebellaris sinistra*	3
9	*Fossa cerebellaris dextra*	3.5
–	*Torus occipitalis* (max.)	16
–	Asterion region (left = right)	14

PHYLOGENY AND TAXONOMY

It seems highly probable that Vértesszöllös Man takes his origin from a relatively near common root with *Pithecanthropus, Sinanthropus, Atlanthropus* (using simple find names), and other so-called *Homo erectus* representatives. It seems probable to me that *Homo* arrived in Europe by successive waves of migration. The shortest route for the early waves was the Pleistocene landbridge connecting Asia Minor and the Balkans (Flint 1971). Tentatively, we can link Petralona, Vértesszöllös, Bilzingsleben, and Mauer to these early waves. According to Henriette Alimen (1975), during the Riss period there was an Acheulean wave of migration from southern routes (Gibraltar, Sicily), which was

TABLE 7.3

Shape distances of fossil occipital bones from a recent reference series by means of biasterionic breadth, lambda-opisthion chord ($l-o_2$), lambda-inion chord, and inion-opisthion chord

Recent Zalavár compared with	$\frac{3}{4}C_z^2$
Oberkassel female	0.10
Oberkassel male	0.28
Skhul V	1.00
Vértesszöllös	1.39
Swanscombe	1.89
La Ferrassie I	2.58
Sinanthropus III	5.31
La Chapelle	5.41
Pithecanthropus IV	12.14

marked by African-type cleavers (cf. Freeman 1975:733). We can identify Steinheim, Swanscombe, Arago, and Biache as some possible members of this more gracile group. Their occipital structure displays features in common with Rabat and Saccopastore. We must suppose that later and by orthoselection this human amalgam evolved into classic neanderthals because, in the Early Würm, the latter completely covered Europe.

In customary morphological systematics, paleospecies like *Homo erectus* are valid taxa. It is clear (as it was in 1966) that the Vértesszöllös II occipital squama is morphologically intermediate between *H. erectus* and *H. sapiens*. However, it is quite possible that the Vértesszöllös II skeleton, in its entirety, might have been unambiguously assignable to one of the two species. To give expression to this situation, Vértesszöllös II has received the taxonomic name, *Homo (erectus seu sapiens) palaeohungaricus* n. ssp. (Thoma 1966:531), which means: genus *Homo*, species *erectus* or *sapiens*, subspecies *palaeohungaricus*.

This solution was suggested by Dr Miklos Kretzoi whose advice I gratefully acknowledge. The combination of characters, which is new for both species, has made necessary the creation of a new subspecies, whose holotype is Vértesszöllös II.

In phylogenetic systematics (Hennig 1966) we must go back in time from the recent taxon (at least a species) to the last bifurcation. This bifurcation generates a sister group (extant or extinct). Sister groups must have equal systematic rank, depending upon the time depth of their separation. According to Robinson (1972a), the last clear-cut bifurcation in the hominid family created (perhaps

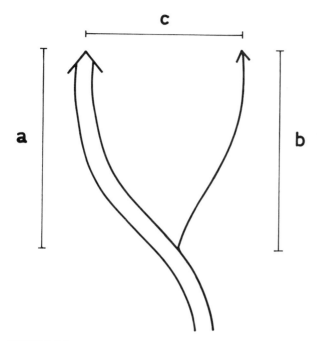

FIGURE 7.6
Intraspecific radiation. Explanation in the text.

three million years ago) two genera, *Paranthropus* and *Homo*. It seems that Robinson's bigeneric classification of Hominidae is the only consequently phylogenetic one. It is analogous to von Huene's (1940) Archosauromorpha group which comprises two sister groups, crocodiles and birds (with proto-birds). The first group remained in the reptilian stage, while the second crossed and surpassed this stage. In any case, the genus *Homo* was monotypic above the australopithecine stage and, therefore, in the phylogenetic system, Vértesszöllös Man is *sapiens*.

There is another argument in favor of the above classification. This is Weidenreich's (1943) evolutionary line, *Pithecanthropus* – Ngandong – Australians (including Kow Swamp). Such an isolated evolution is theoretically possible. Calculations using Malécot's discontinuous, bidimensional migration model demonstrate powerful isolation by distance in Paleolithic conditions (Thoma 1976c). If in Figure 7.6, the right evolutionary line is that of Weidenreich and the left one lumps the rest of mankind, the genetic difference c

results from accumulated quantities of point mutations and chromosomal changes a and b in the two lines. Consequently, $c > a, b$. Because c does not indicate a species difference, neither does a or b. Therefore, the boundaries of the species *H. sapiens* must be extended back in time, at least to the common ancestor. Of course, Vértesszöllös Man is included.

So what happens to *Homo erectus*? In my opinion, for the Javanese pithecanthropines, Ngandong Man, and some fossil and subfossil Australians, we have to create a new subspecies, *Homo sapiens erectus*.

CONCLUSIONS

In the customary morphological systematics, the fragmentary Vértesszöllös finds seem to be intermediate between *Homo erectus* and *Homo sapiens*. In the phylogenetic (Hennig 1966) system, Vértesszöllös Man belongs to the species *Homo sapiens*.

8

MARIE-ANTOINETTE DE LUMLEY

Les Anténéandertaliens en Europe

The European Archanthropines are referred to as Anteneanderthals because they lived before Neanderthal Man, between 800,000 and 120,000 years ago. Numerous recent discoveries have given us a better understanding of the physical appearance of the most ancient men in Europe and, more particularly, in the south of France. The skull discovered at the Caune de l'Arago in the eastern Pyrenees represents, with the one from Steinheim in Germany, one of the two oldest known skulls in Europe.

De nombreuses découvertes récentes permettent actuellement d'avoir une certaine connaissance du premier peuplement de l'Europe: c'est-à-dire de repérer dans le temps les premières manifestations de la présence humaine sur ce continent et de préciser quel était cet Homme qui vivait avant l'homme de Néandertal.

LES HOMMES DU PLÉISTOCÈNE INFÉRIEUR

Si les plus vieux outils taillés par l'Homme ont été datés en Afrique orientale (Kenya, Ethiopie) d'environ trois millions d'années, l'arrivée de l'Homme en Europe paraît beaucoup plus récente. En effet, les plus anciens sites où des outils taillés par l'Homme préhistorique ont été recueillis, peuvent être datés en Europe de 1,800,000 ans pour Chilhac dans le Massif Central (France), de 1,000,000 pour Cullar de Baza I près de Grenade (Espagne), de 950,000 à 900,000 ans pour la Grotte du Vallonnet à Roquebrune-Cap-Martin (Alpes-Maritimes, France). Plus de 100 stations sur les hautes terrasses des fleuves côtiers du Roussillon dans le Sud de la France étaient parcourues par les hommes préhistoriques au Pléistocène inférieur.

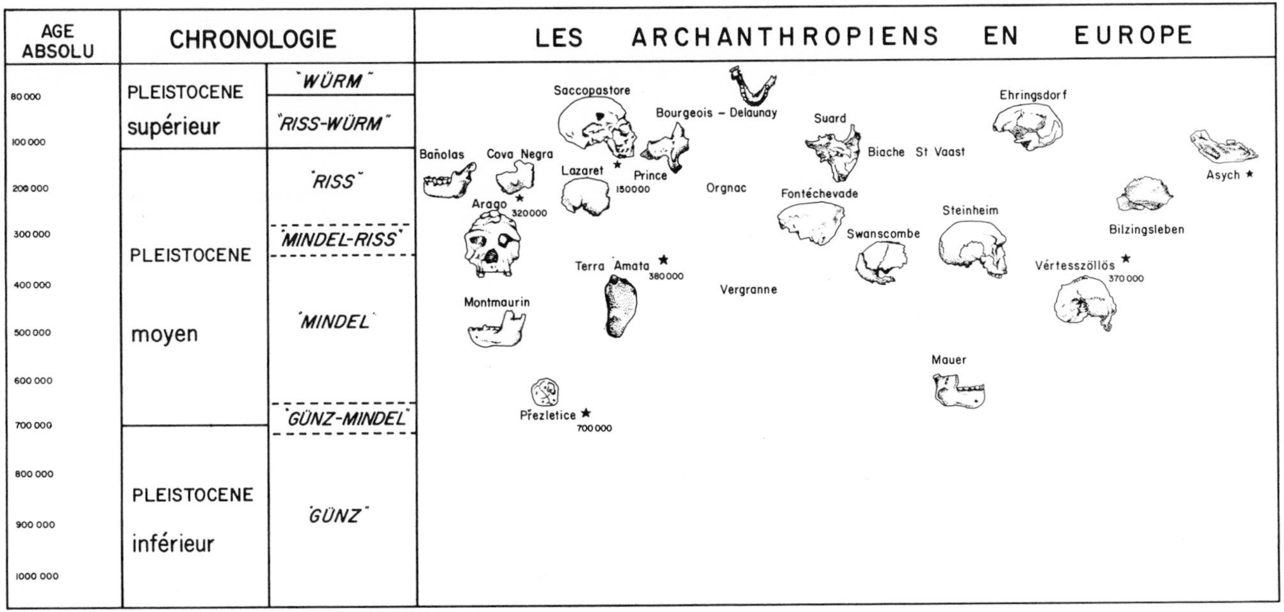

FIGURE 8.1

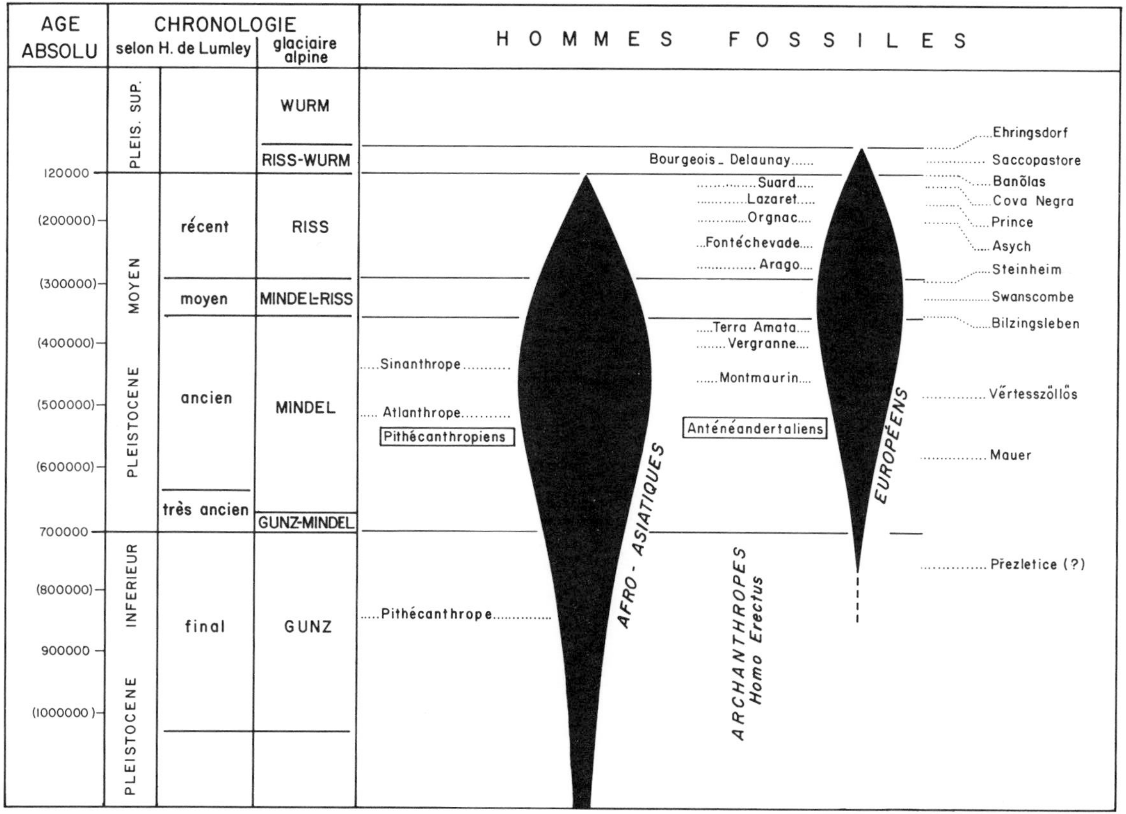
AGE ABSOLU
CHRONOLOGIE
selon H. de Lumley
glaciaire alpine
HOMMES FOSSILES
120000
(200000)
(300000)
(400000)
(500000)
(600000)
700000
(800000)
900000
(1000000)
PLEIS. SUP.
PLEISTOCENE MOYEN
PLEISTOCENE INFERIEUR
récent
moyen
ancien
très ancien
final
WURM
RISS-WURM
RISS
MINDEL-RISS
MINDEL
GUNZ-MINDEL
GUNZ
Sinanthrope
Atlanthrope
Pithécanthropiens
Pithécanthrope
AFRO - ASIATIQUES
Bourgeois_ Delaunay
Suard
Lazaret
Orgnac
Fontéchevade
Arago
Terra Amata
Vergranne
Montmaurin
Anténéandertaliens
ARCHANTHROPES
Homo Erectus
EUROPÉENS
Ehringsdorf
Saccopastore
Banõlas
Cova Negra
Prince
Asych
Steinheim
Swanscombe
Bilzingsleben
Vértesszöllös
Mauer
Přezletice (?)

FIGURE 8.2

Néanmoins, si la présence d'outils taillés permet d'affirmer l'existence de l'Homme en Europe depuis plus d'un million et demi d'années aucun reste osseux humain, antérieur à 800,000 ans, n'a été découvert jusqu'à ce jour rien ne peut faire connaître l'aspect physique de cet Homme.

Ne paraissant pas connaître le feu, il vivait soit en plein air, où les campements sont bien localisés par la concentration des outils découverts, soit en grotte. L'organisation de l'habitat était des plus élémentaires et les techniques de chasse qui peuvent être étudiées au Vallonnet, par exemple, devaient être rudimentaires, si l'on considère la forte proportion d'animaux âgés, vraisemblablement des bêtes malades ou mortes, ramenées dans l'habitat.

LES HOMMES DU PLÉISTOCÈNE MOYEN TRÈS ANCIEN
(800,000 A 500,000 ANS)

Un fragment de molaire recueillie en Tchécoslovaquie à Prezletice et une mandibule en Allemagne à Mauer, représentent les plus anciens et les seuls témoins osseux des Hommes fossiles connus actuellement en Europe: les Archanthropiens européens. A cette appellation nous préférons celle d'Anténéandertaliens. Prise essentiellement dans un sens chronologique et non phylétique, elle désigne la population qui vivait en Europe avant l'Homme de Néandertal entre 800,000 et 80,000 ans.

LES HOMMES DU PLÉISTOCÈNE MOYEN ANCIEN
(500,000 A 300,000 ANS)

Les hommes sont encore peu nombreux dans toute l'Europe.

En Hongrie, à Vértesszöllös, un occipital d'adulte et des dents d'enfants, ont été recueillis dans un ensemble stratigraphique bien daté.

En France, une mandibule isolée est découverte en 1949 par R. Cammas dans la grotte de la Niche à Montmaurin dans la Haute-Garonne. Son âge géologique est imprécis: Mindélien ou Mindel-Riss. Elle offre quelques ressemblances avec les mandibules de Mauer et de l'Arago: il ne semble pas douteux d'après sa morphologie qu'elle se place dans le groups des Anténéandertaliens.

Ses dimensions, moins robustes que celles des mandibules de Mauer et Arago XIII, seraient en faveur de son appartenance au sexe féminin, comme Arago II.

Elle présente une grande largeur comparable à la mandibule de Mauer. L'angle symphysien élevé traduit une importante fuite du menton, comme sur les mandibules Arago II et XIII.

Sur la face interne, la présence d'un planum alvéolaire paraît être un caractère propre aux mandibules anténéandertaliennes. D'autre part la disposition de la ligne oblique interne est assez particulière; elle est très différente de celle des hommes modernes et de celle des Néandertaliens. Elle reste en position haute, sur la mandibule de Montmaurin et rejoint en avant le bord supérieur de la fosse génio-glosse où elle se raccorde avec le margo-terminalis. Chez l'Homme moderne, la ligne oblique interne suit un trajet très oblique en bas et en avant.

Les empreintes digastriques, situées entièrement sur le bord inférieur, sont très vastes et s'étendent jusqu'au niveau des premières molaires. La branche montante est large et l'échancrure sigmoide peu profonde.

Les molaires, peu usées, devaient appartenir à un jeune adulte; croissantes de M1 à M3, elles présentent un schéma dryopithécien.

Dans le Doubs, une canine humaine appartenant vraisemblablement à un enfant est recueillie en 1973 par Michel Campy dans l'aven de Vergranne. Bien qu'isolée de tout contexte archéologique, cette dent déciduale est associée à une faune d'âge Mindel récent. D'autre part, ses grandes dimensions et sa forme globuleuse rappellent la morphologie des dents de l'Arago.

Sur le site de Terra Amata à Nice, au cours de la campagne de fouilles d'urgence effectuée de janvier à juillet 1966 par H. de Lumley et son équipe, une empreinte de pied humain est dégagée sur un des sols d'habitats acheuléens anciens. Cette empreinte de 24 cm de longueur paraît correspondre à un individu qui pouvait mesurer un peu moins d'un mètre soixante, sans préciser toutefois s'il s'agit d'un adolescent ou d'un adulte.

Ces hommes découvrent l'usage du feu. Les plus anciens foyers aménagés peuvent être datés de cette période: Terra-Amata et Lunel-Viel en France, Vértesszöllös en Hongrie, Torralba et Ambrona en Espagne, Torre in Pietra en Italie.

D'autre part, les premiers campements organisés en plein air, ou en grotte, font leur apparition. Sur les sols d'habitat de Terra Amata, il est possible de distinguer des ateliers de taille, des foyers creusés dans des cuvettes de sable, et protégés par une murette de pierres. Ces foyers sont avec ceux de Vértesszöllös les plus anciens foyers aménagés connus dans le monde.

Les techniques de chasse se développent, avec la vie sociale. Les hommes s'attaquent le plus souvent à des animaux jeunes, plus faciles à abbattre (éléphanteaux à Terra Amata).

LES HOMMES DU PLÉISTOCÈNE MOYEN MOYEN ET MOYEN RÉCENT (300,000 À 120,000 ANS)

Les restes humains sont un peu plus nombreux. C'est en Europe occidentale qu'il est possible de percevoir quelques chaînons de l'évolution humaine. En effet, grâce à des découvertes récentes une certaine connaissance de ces hommes est désormais permise.

Aucun squelette complet ni aucune trace de sépulture n'ont été retrouvés sur les sols d'habitats de ces Archanthropiens européens ou Anténéandertaliens.

Seuls des os isolés et fragmentés ont été recueillis, les plus souvent mêlés aux ossements d'animaux chassés et évoquant la pratique d'une anthropophagie.

Ces ossements humains, même fragmentaires, restent cependant d'un intérêt capital et une analyse anatomique détaillée apporte d'intéressantes données sur ce que fut l'aspect physique de ces hommes et sur leur signification phylétique.

Des restes humains ont été recueillis dans la Caune de l'Arago, la grotte du Lazaret, Orgnac, la grotte du Prince, Fontéchevade, Suard, Cova Negra, Atapuerca, et Bañolas.

Les Restes Humains de la Caune de l'Arago

Dans la grotte ou 'Caune de l'Arago' située au nord de la plaine du Roussillon sur la commune de Tautavel à 19 km au nord-ouest de Perpignan (Pyrénées-Orientales), d'importantes campagnes de fouilles sont organisées par Henry de Lumley depuis 1964.

Sur les différents sols d'habitats préhistoriques remontant à l'extrême début du Rissien, ont été découverts des restes humains isolés, dispersés au milieu d'un amoncellement d'outillage en pierre tayacien et d'ossements d'animaux chassés par les groupes de nomades préhistoriques qui revenaient régulièrement établir leur campement provisoire dans cette grotte.

De nombreux restes humains ont été recueillis: une vingtaine de dents isolées, des fragments de pariétaux, une rotule, des phalanges et en particulier une mandibule avec six dents en place (Arago II, en juillet 1969) une demi-mandibule portant cinq dents (Arago XIII, en juillet 1970), et la partie antérieure d'un crâne d'adulte (Arago XXI, le 22 juillet 1971).

La taille très différente des deux mandibules paraît traduire un important dimorphisme sexuel: Arago XIII ayant appartenu à un sujet de sexe masculin d'environ 20 ans et Arago II à un individu féminin d'environ 40 à 55 ans.

Ces mandibules présentent des caractères archaiques qui les rapprochent de celles de Mauer et de Montmaurin. Leurs dimensions sont grandes: la largeur de la mandibule (diamètre bicondylien externe) est, en particulier pour Arago XIII, très élevée. L'arcade alvéolaire construite au-dessus de l'arcade basilaire, est convexe en avant; la symphyse est fuyante; le triangle mentionnier absent, les empreintes digastriques sont entièrement situées sur le bord inférieur de l'os, la fosse génioglosse est profonde, le *torus transversus inferior* et le bourrelet marginal sont très épais, la proéminence latérale est forte.

Mais ce qui frappe c'est, en particulier sur Arago XIII, le grand développement du planum alvéolaire, de la ligne oblique interne, la position très basse du trou géni-supérieur et des trous mentionniers.

La branche montante est haute et très large: la condyle a de grandes dimensions; l'apophyse coronoide est courte, basse et large; l'échancrure sigmoide est faiblement marquée, et son point le plus bas est situé nettement en arrière. La crête pharyngienne est très saillante et le bourrelet triangulaire puissant. Le triangle rétro-molaire est vaste; la gouttière rétro-molaire est large et profonde.

Arago II a une branche montante verticale qui rappelle celle de la mandibule de Mauer. Sur Arago XIII, elle est légèrement inclinée en arrière comme sur la mandibule de Montmaurin. La disposition du composant alvéolaire de la mandibule rappelle sur Arago II celle du fossile de Montmaurin et sur Arago XIII celle de Mauer. Une dépression mandibulaire mettant en évidence un léger menton osseux, peut être observée sur Arago II: elle est absente sur Arago XIII comme sur Mauer et Montmaurin.

Les dimensions des dents de ces deux mandibules sont très différentes. Les unes de taille moyenne appartiennent à la mandibule féminine (Arago II), les autres sont de très forte taille, dépassant les dimensions relevées sur les fossiles de Mauer et de Montmaurin et appartiennent à la mandibule du type masculin (Arago XIII).

Le même dimorphisme sexuel existe au niveau des dents isolées correspondant à des individus d'âge différent: enfants, adultes. Il est intéressant de noter que le schéma formé par les cuspides des molaires n'est pas particulièrement archaique et la troisième molaire est légèrement plus petite que la seconde.

Par l'ensemble de leurs dimensions et leur morphologie, les mandibules de l'Arago présentent une certaine homogénéité avec les autres mandibules anténéandertaliennes d'Europe: Mauer, Montmaurin, Asych, et Bañolas. Par bien des caractères, par contre, elles paraissent se distinguer de celles des Archanthropiens qui vivaient de l'autre côté de la Méditerranée: en Afrique et dans le Sud-Est asiatique, Atlanthropes, Pithécanthropes, Sinanthropes.

Le crâne (Arago XXI) découvert en juillet 1971, gisait renversé sur le sol d'habitat, posé sur la calotte, la voûte palatine en l'air. Il avait été déposé ou abandonné au milieu d'ossements de rhinocéros, de chevaux, de boeufs, de cerfs, de bouquetins, et au même niveau se trouvaient de nombreux outils appartenant à une industrie tayacienne ou acheuléenne.

Ce crâne, comprenant essentiellement le frontal, le sphénoide, l'ethmoide et la face complète (maxillaires supérieurs, vomer, malaires, palatins, unguis, os propres du nez), devait appartenir à un individu âgé d'une vingtaine d'années d'après la faible trace d'usure de la troisième molaire en place sur l'arcade dentaire, et l'absence de synostose de la suture coronale de l'os frontal.

Ainsi la seule face connue de l'Homme rissien en France est celle de l'Arago. Elle peut être comparée avec celle de l'Homme de Steinheim. Les indices de courbures mesurés sur le frontal mettent en évidence une platicranie comparable ou plus forte que celle de la plupart des Néandertaliens. Le front n'est cependant

122 *Homo erectus*

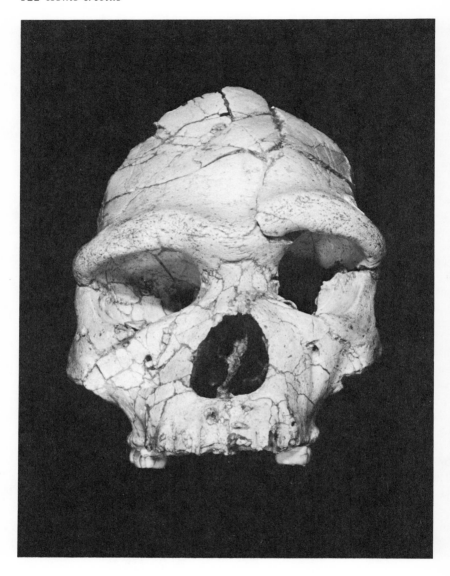

FIGURE 8.3
Arago XXI. Crâne Anténéandertalien découvert dans la Caune de l'Arago, à Tautavel dans le sud de la France, le 22 juillet 1971 (grandeur: nature).

FIGURE 8.4
Arago XXI. Vue latérale gauche du crâne Anténéandertalien découvert dans la Caune de l'Arago à Tautavel.

pas aussi plat que chez certains Archanthropiens classiques — Pithécanthropes, Sinanthropes — et il n'y a pas de carène sagittale.

L'examen des proportions relatives du frontal de l'Homme de Tautavel montre un développement dans le sens longitudinal plus grand que chez la plupart des Néandertaliens; par contre il paraît plus petit dans le sens transversal par suite d'un fort rétrécissement post-orbitaire.

En vue supérieure, ce rétrécissement post-orbitaire est bien visible. Il est nettement plus fort que sur les crânes des Néandertaliens et paraît comparable à celui de Steinheim; il reste cependant moins important que sur les Sinanthropes et les Pithécanthropes. Alors que, contrairement à l'homme moderne, les bosses frontales sont effacées chez les Néandertaliens adultes, elles ne sont même pas repérables sur le frontal de l'homme de Tautavel comme sur celui de Steinheim.

Les crêtes latérales du frontal sont saillantes et très hautes. Les facettes temporales du frontal sont vastes et plus profondes que sur les crânes de Steinheim et des Néandertaliens.

Le torus sus-orbitaire est très fort. On peut distinguer l'arcade superciliaire et l'arcade supra-orbitaire. Elles sont en partie séparées par un sillon oblique, vers le haut, et du côté externe: le sulcus supra-orbitalis. Le trigone supra-orbitaire n'est pas individualisé de l'arcade supra-orbitaire. Le torus a sa section la plus forte à environ 30 mm du plan sagittal comme chez l'homme de Steinheim.

La glabelle est déprimée par rapport au torus comme chez l'homme de Steinheim. Le torus ne constitue donc pas un bourrelet continu comme chez les Néandertaliens, mais il est très proche de celui que l'on peut observer sur l'homme de Steinheim. La fosse supratoriale est peu profonde. Elle est moins marquée sur le frontal de Steinheim. La fosse supraglabellaire est bien individualisée.

Les sinus frontaux visibles à la radiographie sont très petits et leur dimension est sans rapport avec le fort développement du torus.

Les orbites sont grandes, basses et rectangulaires, beaucoup moins hautes que chez les Néandertaliens classiques; leur ouverture évoque celles de l'homme de Steinheim. L'écartement interorbitaire est extrêmement important; il est comparable ou légèrement supérieur aux valeurs les plus fortes mesurées chez les Néandertaliens.

Le prognathisme est marqué, la face se projette nettement en avant du crâne cérébral. Cet aspect est cependant exagéré par un fort prognathisme alvéolaire. Les maxillaires supérieurs ont un aspect massif et robuste, leur face antérieure à peu près plane, est dépourvue de fosse canine, comme chez les Néandertaliens. Rappelons cependant qu'une légère dépression est décelable chez l'homme de la Ferrassie. Chez l'homme de Steinheim, la fosse canine est bien marquée et très profonde. Chez les Archanthropiens classiques, elle existe semble-t-il chez le

Sinanthrope, alors qu'elle est absente sur le maxillaire supérieur du Pithécanthrope IV.

Le palais est profond. L'arcade alvéolaire a une forme à tendance upsiloide, les alvéoles des 2P et des 3M peuvent être placés sur une même droite. Cet homme avait perdu de son vivant ses deux deuxièmes prémolaires et leurs alvéoles sont plus ou moins résorbés. M2 est la dent la plus grosse et M3 est relativement réduite: M1 < M2 > M3. La face occlusale des molaires est peu compliquée et il n'existe pas de cingulum.

Un bourrelet mandibulaire externe très saillant, qui se confond avec le bord alvéolaire externe, est bien visible au niveau des alvéoles de M1 à M3. Un bourrelet semblable peut être observé sur l'homme de Steinheim, mais il existe aussi chez certains Néandertaliens, la Ferrassie par exemple.

La saillie du bec encéphalique devait être particulièrement développée, en rapport avec le fort écartement des orbites. Un premier examen de l'endocrâne montre que la troisième circonvolution frontale était bien développée et le cap de Broca devait être très saillant. Le volume endocrânien devait être relativement petit, certainement plus faible que celui de la moyenne des Néandertaliens classiques mais un peu plus fort que chez l'homme de Steinheim.

Les Restes Humains de la Grotte du Lazaret
La grotte du Lazaret, située à Nice à l'est du port de commerce, s'ouvre sur les pentes occidentales du Mont Boron tout près du rivage actuel de la Méditerranée. Connue depuis fort longtemps, elle fut l'objet de patientes et minutieuses recherches du Commandant Octobon et de Noëlle Chochon. En 1953, 1958, et 1964, ils découvrirent quelques restes humains deux dents (une incisive de lait et une canine d'adulte) et un pariétal d'enfant. La position stratigraphique de ces trois éléments est parfaitement bien établie; ils proviennent de la base des niveaux du Riss III du Locus VIII.

Ces restes humains correspondent à trois individus différents, deux enfants âgés d'environ deux ans et neuf ans et un adulte.

L'incisive temporaire (il sup. g.) se différencie des dents modernes par de très grandes dimensions, par un bombement marqué de la couronne du côté vestibulaire et une forme en 'pelle' avec une éminence cervicale du côté lingual.

La canine d'adulte (C. inf. d.) de grande taille, a des dimensions plus grandes que celles relevées sur les mandibules de Mauer et de l'Arago; elles dépassent les valeurs moyennes calculées chez les Néandertaliens. Le modelé de la couronne se caractérise par une face linguale en forme de pelle et une face vestibulaire régulièrement et faiblement convexe, sans renflement mésio-cervical.

Le pariétal droit de l'enfant de Lazaret, en bon état de conservation, est aplati et rectangulaire. L'indice d'allongement très faible 81.2 met en évidence un

126 *Homo erectus*

pariétal plus allongé que celui de tous les hommes fossiles. D'autre part, le bord temporal est légèrement plus long que le bord sagittal contrairement à ce qui peut être observé chez l'homme moderne. Cette disposition exceptionnelle paraît traduire une bascule de l'occipital peu développée. Le degré d'élévation du pariétal peut donner une idée approximative du degré de cérébralisation. Il est légèrement supérieur à celui des Archanthropiens d'Afrique et d'Asie, mais reste inférieur à celui des Néandertaliens. Le dessin de l'artère méningée, bien visible sur la face endocrânienne, peut être considéré dans son ensemble comme une variante du type humain moderne.

Cependant, la complexité du rameau lambdatique représente une persistance d'une disposition primitive en rapport avec l'importance relativement restreinte de la région cérébrale antérieure chez cet Anténéandertalien.

Le pariétal du Lazaret présente une vaste lésion. Il s'agit d'un remaniement osseux de type à la fois lacunaire et hypertrophiant qui intéresse la corticale externe et interne de l'os. De nombreux pertuis vasculaires endocrâniens disposés en 'pomme d'arrosoir,' et une modification du trajet des rameaux de l'artère méningée moyenne sont en faveur d'une étiologie tumorale. Il pourrait s'agir d'une tumeur des méninges et cet enfant serait vraisemblablement mort d'un méningiome.

Les Restes Humains d'Orgnac
Les recherches effectuées en Ardèche, par Jean Combier et son équipe dans le site d'Orgnac III, ont permis de découvrir dans les niveaux inférieurs du Rissien (niveaux 5a, 5b, 6), associées à une industrie acheuléenne supérieure, des dents humaines bien conservées: une canine supérieure droit en 1962 et six dents temporaires; quatre molaires et deux incisives de 1968 à 1971. Seule la canine devait appartenir à un adulte. Parmi les six dents de lait, deux molaires (Homo 2 et 9) devaient appartenir à un enfant âgé de neuf ans et deux incisives (Homo 7 et 8), à un enfant plus jeune, âgé de cinq ans environ.

L'une des molaires de lait (*Homo* 2) est une m2g inférieure dont l'usure de la couronne et l'état d'avancement de la résorption de la racine permettent de l'attribuer à un enfant âgé de neuf ans environ. La couronne de cette dent a des dimensions élevées, supérieures à celles d'un enfant actuel et à celles des enfants néandertaliens. D'autre part, sa forme allongée et rectangulaire lui est propre; en effet, les dents modernes sont plutôt carrées. Néanmoins, le nombre des cuspides (5) ne diffère pas de celui des hommes actuels. Cette dent d'Orgnac présente une ressemblance extraordinaire avec les deux secondes molaires inférieures de lait des enfants anténéandertaliens de l'Arago (Arago I et V), et leurs traits communs sont essentiellement: un allongement mésio-distal important de la couronne, une fovea antérieure grande et profonde, et un bourrelet cervical développé du côté vestibulaire.

Enfin, il est intéressant de noter que le site d'Orgnac III présente de grandes analogies avec la grotte du Lazaret au point de vue stratigraphique et archéologique. Les hommes qui vivaient à Orgnac III étaient vraisemblablement contemporains de ceux qui vivaient au Lazaret.

Les Restes Humains de la Grotte du Prince
A 100 mètres de la frontière italienne à Grimaldi, de nouvelles recherches entreprises par L. Barral et S. Simone dans la grotte du Prince ont permis de dégager en 1968, dans des niveaux rissiens, un os iliaque droit d'une femme adulte. C'est le premier os iliaque anténéandertalien connu actuellement en Europe. Bien qu'incomplet il présente une structure générale et des empreintes musculaires très proches de celles des iliaques modernes, ce qui confirme que les Anténéandertaliens, comme les Néandertaliens avaient une station droite et parfaitement redressée.

Les différences observées par rapport aux os modernes peuvent être considérées comme des caractères de détail dont l'intérêt réside dans le fait qu'elles sont presque toujours retrouvées sur les os iliaques néandertaliens: une double courbure de l'aile iliaque atténuée dont la portion antérieure est réduite, présence d'une fossette sus-cotyloidienne, une cavité cotyloide ovalaire, une gouttière du psoas-iliaque très profonde, une échancrure interépineuse très étroite, une épine iliaque antéro-inférieure allongée, déjetée, bien individualisée. Tous ces caractères existent sur les os des Néandertaliens de la Chapelle-aux-Saints, de la Ferrassie, de Krapina et de l'Hortus.

Par contre, une faible profondeur de la cavité cotyloide et une surface rétro-cotyloidienne très étroite paraissent constituer des caractères propres à l'Anténéandertalien de la grotte du Prince.

Les Restes Humains de Fontéchevade
En 1947, Germaine Henri-Martin mettait au jour dans la grotte de Fontéchevade, proche de la petite ville de Montbron, quelques restes humains: une portion de calotte crânienne (Homo II) comprenant la majeure partie des pariétaux et du frontal, et un petit fragment de frontal (Homo I), associés à une industrie tayacienne.

Une récente étude sédimentologique du gisement permet à A. Debénath de reconsidérer la stratigraphie et d'attribuer un âge plus ancien à la calotte crânienne. La couche tayacienne dans laquelle elle a été découverte n'appartiendrait pas au Riss-Würm, mais à une phase relativement ancienne de la glaciation rissienne, peut-être le Riss II.

La calotte crânienne est épaisse et aplatie. Les bosses pariétales sont à peine marquées et les lignes temporales, très effacées, sont situées sur la portion inférieure de l'os. Le réseau artériel méningé a laissé peu d'empreintes sur la face endocrânienne du pariétal. Le rameau obélien est le plus développé et le rameau bregmatique antérieur est réduit comme chez les autres Anténéandertaliens.

Les Restes Humains de la Grotte Suard
Ce site connu depuis plus d'un siècle (1870) fut l'objet de nombreuses fouilles sporadiques. Plusieurs restes humains d'enfants et d'adultes, généralement fragmentaires, ont été recueillis par F. Bordes, P. David, et A. Debénath. Une récente analyse approfondie de la faune, de l'industrie, ainsi que l'étude sédimentologique conduisent à ranger dans le Rissien supérieur les divers niveaux de la grotte Suard.

Pour Jean Piveteau, le bombement de la région pariétale est supérieur à celui de l'Atlanthrope et des divers Sinanthropes, mais reste en-dessous de celui des Néandertaliens. Par d'autres traits, ces Anténéandertaliens de Suard, annoncent les populations néandertaliennes: artère méningée moyenne, avec une partie antérieure prédominante, morphologie interne et externe de l'occipital, certains caractères du temporal.

Le Pariétal de Cova Negra
Un pariétal droit humain a été découvert le 11 juillet 1933 par le Père Gonzalo Vines, dans la couche moyenne du remplissage quaternaire de la grotte de Cova Negra (Jativa, province de Valence en Espagne).

L'examen de l'industrie par Henry de Lumley montre qu'il s'agit d'une industrie proto-charentienne ou tayacienne de même type que celle décrite à la Micoque (Dordogne), à la Caune de l'Arago (Pyrénées-Orientales), à la Baume Bonne (Basses-Alpes) et datée de l'époque rissienne. La faune archaique est représentée par: *Elephas antiquus, Rhinoceros mercki, Equus caballus mosbachensis.*

Assez bien conservé, ce pariétal devait appartenir à un individu adulte qui, d'après l'état d'oblitération des sutures, pouvait avoir une cinquantaine d'années. Les grandes dimensions, la forte épaisseur de l'os, les saillies bien marquées des insertions musculaires sont en faveur du sexe masculin.

Dans une précédente étude M Fusté avait attribué ce pariétal au type néandertalien. En effet, certains caractères pourraient le rapprocher des Néandertaliens: la platycrânie antéropostérieure, le contour de la voûte en vue postérieure régulièrement arrondi, la bosse pariétale peu marquée en position inférieure et reculée, les lignes temporales dont le relief est assez net, relativement basses, enfin la présence d'une dépression paralambdoide.

Une nouvelle étude plus détaillée de ce pariétal nous a permis d'observer d'autres caractères qui le différencient nettement des Néandertaliens, et le rapproche des Anténéandertaliens.

Il apparaît en effet, par certains caractères moins évolué que les Néandertaliens:

1 L'indice qui compare le diamètre bistéphanique à la largeur bipariétale maximale est en effet inférieur aux valeurs obtenues chez les Néandertaliens, et témoigne d'un fort rétrécissement de la région frontale.
2 Le biseau de la suture temporale est court et bien délimité. Une disposition semblable a été décrite par H.V. Vallois sur les pariétaux rissiens de Fontéchevade et de Swanscombe, tandis que chez les Néandertaliens et chez l'homme actuel, ce biseau est mal délimité et plus long.
3 La disposition de l'angle postéro-inférieur rappelle celle observée sur le pariétal de Swanscombe, où astérion et métastérion sont présents suivant un type intermédiaire à celui signalé, d'une part chez les Néandertaliens et d'autre part, chez l'homme moderne.
4 L'épaisseur de l'os dans cette région astérique est considérable et s'explique par l'absence sur la face endocrânienne, de la gouttière du sinus latéral, habituellement visible sur le pariétal des Néandertaliens et de l'homme actuel.
5 Les empreintes vasculaires de l'artère méningée moyenne offrent un certain aspect archaique, par le développement relativement grand du réseau lambdatique et par la situation du rameau bregmatique entièrement compris sur la moitié antérieure du pariétal.
6 L'encéphale avait une scissure de Sylvius large et des circonvolutions pariétales ascendante et supérieure étroites.

A ces éléments, s'ajoutent de nombreux caractères plus évolués que ceux observés chez les Néandertaliens. L'homme de Cova Negra se caractérise par:

1 Un pariétal relativement court par rapport à son exceptionnelle largeur (152 mm) dont la valeur dépasse celle relevée aussi bien sur les Néandertaliens que sur l'homme actuel.
2 Un astérion en position inférieure, rapprochée vers l'avant qui indiquerait une rotation marquée de l'occipital.
3 Une hauteur maximale plus élevée que celle des Néandertaliens dont la valeur est située à l'intérieur des marges de variations de l'homme actuel.
4 Une forte courbure de l'os dans le sens transversal traduisant un enroulement du pariétal comparable à celui de l'homme moderne.
5 Une disposition en éventail, ouvert vers l'arrière, des branches de l'artère méningée.

Le pariétal de Cova Negra serait donc par certains caractères plus évolué que celui des Néandertaliens. Ces observations sont confirmées par l'étude des sphères péribipariétales dans lesquelles s'inscrivent les pariétaux par leurs points bregma, lambda, fronto-pariétal. A. Delattre et R. Fenart ont montré que le phénomène de sphérisation ontogénique de la voûte bipariétale humaine a entraîné la mise en place sur cette sphère de deux autres points, astérion et point incisural interne.

Cette disposition est réalisée chez les Néandertaliens comme chez l'homme moderne. Nous l'avons retrouvée sur le pariétal rissien de Cova Negra, dont la sphérisation était achevée. Il n'en est pas de même sur le pariétal du Lazaret, qui lui aussi date du Riss, mais qui ne paraît pas avoir atteint le stade évolutif du fossile qui fait l'objet de cette étude.

Il semble donc, qu'à l'époque rissienne, deux types humains anténéandertaliens vivaient sur les côtes méditerranéennes: (1) l'homme du Lazaret à qui peut être associé l'homme de la Chaise, et (2) l'homme de Cova Negra à qui peuvent être associés les fossiles de l'Arago, de Fontéchevade et de Swanscombe. Ces différents types humains traduisent l'existence d'un grand polymorphisme au sein des Anténéandertaliens.

Les Restes Humains d'Atapuerca
Une récente découverte, effectuée en août 1976, a permis de recueillir dans la grotte d'Atapuerca, près de Burgos, en Espagne, deux mandibules, treize dents isolées et deux fragments de pariétaux, qui représentent un total de cinq à huit individus.

Les mandibules d'Atapuerca, associées à une faune du Pléistocène moyen, présentent de nombreux caractères communs avec les mandibules anténéandertaliennes européennes: Arago II, Arago XIII, Mauer, Montmaurin, et Banolas.

La forme générale de la mandibule est longue, large, basse, et très épaisse. La région symphysienne est la plus caractéristique, avec un angle symphysien ou mentonnier très fuyant, une face antérieure absolument plate, sans relief, et une face postérieure avec un planum alvéolaire très développé. Cette disposition paraît constante chez les Anténéandertaliens européens.

La Mandibule de Bañolas
C'est dans une carrière de travertins à Bañolas (Espagne), qu'une mandibule humaine fut découverte en avril 1887. Souvent datée du Würm, la terrasse d'où provient la mandibule pourrait être vieillie et rapportée au Riss-Würm. Bien conservée, cette mandibule devait appartenir à un individu âgé d'un peu plus d'une cinquantaine d'années et vraisemblablement de sexe féminin. Par sa position stratigraphique, ce fossile antérieur aux Néandertaliens, présente toute une gamme de caractères, à la fois archaiques et plus ou moins évolués.

La mandibule de Bañolas peut être rapprochée des Anténéandertaliens d'Europe occidentale, de Mauer, Arago, Montmaurin par:

1. Une grande longueur, qui est légèrement plus faible que celle de Mauer et Arago XIII;
2. Une grande largeur;
3. Un fort indice de robusticité, qui dépend de la faible hauteur du corps plus que de son épaisseur relativement peu élevée;
4. Une arcade basilaire allongée;
5. Une arcade alvéolaire régulièrement convexe en avant;
6. Une région symphysienne fuyante;
7. L'absence de menton osseux et de triangle mentonnier;
8. Des éminences canines à peine visibles;
9. Une fosse génio-glosse profonde;
10. Une ligne oblique interne très marquée et très bas située; et
11. Une branche montante verticale haute et large.

Elle présente aussi quelques caractères relevés chez les Néandertaliens, en particulier:

1. Une situation de l'arcade basilaire sous l'arcade alvéolaire, qui traduit une fuite de la symphyse moins grande que chez les autres Anténéandertaliens;
2. Une dominance de la canine sur le groupe dentaire canin.

Elle paraît enfin, par certains éléments, plus évoluée que les Néandertaliens:

1. Absence de planum alvéolaire à la face postérieure de la région symphysienne;
2. Position haute du trou géni supérieur;
3. Un *margo terminalis* peu saillant;
4. Situation des empreintes digastriques sur la face postérieure de la symphyse.

Il est évident que tous ces caractères ne présentent pas forcément une valeur phylogénique. La plupart dépendent du développement relatif musculaire et du sexe de l'individu. Mais il reste difficile de discerner l'influence de l'un ou l'autre de ces facteurs devant la rareté actuelle des restes humains fossiles.

Les hommes du Pléistocène moyen récent ont des habitations de plus en plus complexes. Les structures internes font alors leur apparition.

Dans la grotte du Lazaret à Nice, par exemple, les fouilles effectuées ont permis de mettre en évidence une vaste cabane de 11 m de longueur sur 3.50 m de largeur, contre la paroi Est de la grotte. Dans cette cabane des blocages de poteaux, un mur coupe-vent, des issues, une cloison intérieure, des foyers, des zones de litières, ont été nettement repérés. Les chasseurs du Lazaret étaient des nomades qui habitaient dans la grotte temporairement.

132 *Homo erectus*

SOMMAIRE

A la lumière de toutes ces observations, les Anténéandertaliens présentent:

1 Des caractères archaiques qui permettent de considérer qu'ils se trouvent à un stade d'évolution proche des Archanthropes (*Homo erectus*) capacité crânienne réduite, aplatissement du frontal et du pariétal, fort prognathisme, torus sus-orbitaire très développé, rétrécissement post-orbitaire marqué, développement plus ou moins grand de la rotation occipitale, vascularisation méningée réduite dans la portion antérieure du cerveau et, sur la mandibule grand développement du planum alvéolaire, de la ligne oblique interne, position très basse du trou géni-supérieur;
2 Des caractères qui les individualisent des Archanthropiens d'Afrique et d'Asie: fosse supratoriale moins marquée, rétrécissement post-orbitaire moins accusé, absence de carène sagittale;
3 Des caractères qui annoncent les Néandertaliens classiques d'Europe occidentale: l'absence de fosse canine sur l'homme de Tautavel.

Il devait exister un grand polymorphisme chez ces populations rissiennes et le développement des différents individus n'était, semble-t-il, pas équivalent.

En conclusion, si par de nombreux caractères les Anténéandertaliens d'Europe paraissent appartenir au même stade évolutif que les Archanthropiens d'Afrique et d'Asie, par d'autres caractères, ils semblent s'en différencier en offrant une certaine homogénéité européenne. Tout semble se passer comme si les Hommes, après avoir quitté leur berceau africain, et ayant peuplé notre continent il y a plus d'un million d'années, avaient évolué indépendamment sur les différents continents.

9

DIETRICH MANIA & EMANUEL VLČEK

Homo erectus in middle Europe: the discovery from Bilzingsleben

Les restes crâniens hominiens qui peuvent être classés dans la catégorie *Homo erectus* sont décrits dans le cadre d'un complexe travertin du Pléistocène moyen, dans la Thuringe du nord, à Bilzingsleben, République démocratique allemande. Le site suggère la présence à plusieurs reprises d'hominiens du Pléistocène moyen près de l'embouchure d'un ruisseau sur la rive d'un lac. En outre, le site livre des objets de pierre, témoins d'une industrie de petits outils, d'éclats de silex et de pierre taillée. On y trouve aussi des outils frustes sur galets, faits de quartz, de quartzite, de pierre calcaire, et de roche cristalline. De nombreux outils spécialisés sont faits avec des bois de cervidés ou des os délibérément fracturés. La technique de fabrication est la même pour les outils en os et en pierre. Parmi les restes de squelettes animaux, on distingue ceux de l'éléphant et du rhinocéros des forêts, ainsi que du cerf, de l'ours, du castor et du bison européens. Par ailleurs, les restes de *Trogontherium* sont dignes d'attention. Les restes végétaux signalent la présence d'une forêt mixte d'arbres et d'arbustes thermophiles originaires d'Europe centrale, et on remarque une forte influence méditerranéenne. Les mollusques, révélateurs de rapports étroits avec la région du bassin méditerranéen et du sud-est européen, y sont aussi présents. Ainsi, le nord de l'Europe centrale connaissait des conditions atmosphériques plus chaudes et plus douces à l'époque d'*Homo erectus* que de nos jours.

Davidson Black described his first hominid remains from the cave at Chou Kou Tien, a large and robust molar, in 1927. At about the same time Adolf Spengler, a private naturalist, discovered an equally robust and primitive-looking molar in the heart of middle Europe, in travertine quarries near the small village of Bilzingsleben (Thuringia). It was found among stone tools and broken animal bones. Although Black's discovery had great repercussions for continuing excavations at the China fossil site, the tooth from Bilzingsleben did not stimulate further scientific activity. With the cessation of quarrying for travertine, the quarry was closed, and the search for fossils came to an end. The tooth itself was lost.

134 *Homo erectus*

A few publications have dealt with material recovered from the site at Bilzingsleben since its discovery in 1908 by the paleontologist Ewald Wüist (Wiegers 1928, 1940; Andree 1939; Toepfer 1960, 1963, 1970). In 1969, geological and paleontological reconnaissance in the old quarries led to the discovery of a new and rich fossil bed (Mania 1974) and initiated an excavation by the Landesmuseum für Vorgeschichte Halle (Saale). From the start of this excavation the hope was to recover hominid remains. The goal was realized in 1974, and additional discoveries followed in 1975 through 1977 (Grimm, Toepfer, and Mania 1974; Mania and Grimm 1974; Mania 1975, 1976, 1977; Grimm and Mania 1976; Mania, Grimm, and Vlček 1976; Mania and Vlček 1977; Vlček 1978, 1979; Vlček and Mania 1977).

GEOGRAPHIC POSITION OF THE SITE

The site, locally known as Steinrinne, is 1 km south of Bilzingsleben, a small town on the northeastern border of Thuringia (35 km north of Erfurt) in the southwest of the German Democratic Republic (Figure 9.1). The region belongs to the northern part of the European mid-mountain zone; 50 km north, the North German lowlands begin. The site consists of a travertine plateau, the highest point of which is 175 m (NN), and has the coordinates 51°16′24″N, 11°3′39″E.

GEOLOGICAL AGE OF THE SITE

The circumstances of the discovery are determined by the positioning of the fossil bed at the base of a complex of riverine, lacustrine, and limestone (travertine) deposits. This complex accumulated in the bottom of a valley of the Wipper River which today is about 30 m above the level of the valley (Figure 9.2). The site belongs to the middle European river area of Elbe and Saale. According to the morphology of the terrace and the stratigraphic position of the north European glacial advances in this area, a distance of 30 m between ground level and the level of the 'Aue' indicates an age that is younger than the Elster glaciation (Elster/Mindel) but older than the Saale glaciation (Saale/Riss). Therefore, the travertine complex including the fossil bed can be placed in the Elster/Mindel–Saale/Riss interglacial, that is, the Holsteinian of north European glacial chronology. This age is supported by other facts including (1) the composition of the fossil fauna and flora from the travertine complex, (2) the construction of the overlying strata, and (3) preliminary dating by physical-chemical methods (i.e. amino-acid racemization by R. Protsch of 230,000 years BP, a date that represents only a minimum age, as it is much too low for the Holsteinian interglacial).

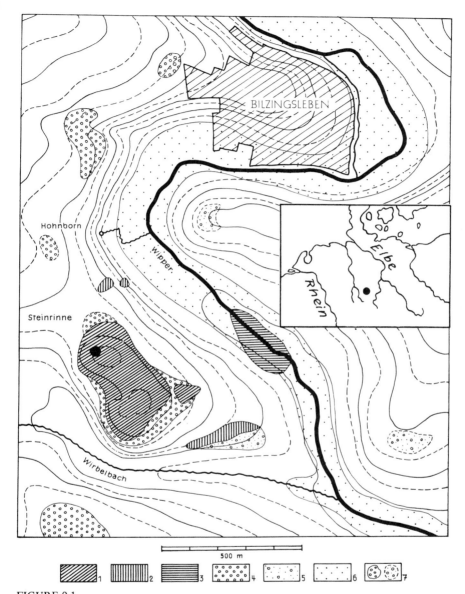

FIGURE 9.1
Morphological and quaternary-geological overview, Bilzingsleben, county of Artern; black circle indicates site of discovery in the quarry: 1, Middle Pleistocene travertine; 2, Upper Pleistocene travertines and limnic formations; 3, Holocene travertines and limnic formations; 4, gravel of the 30 m terrace; 5, gravel of the 20 m and 10 m terraces; 6, Holocene loams on the lower terrace; 7, solid line indicates gravel accumulations; and dotted line gravel distributions.

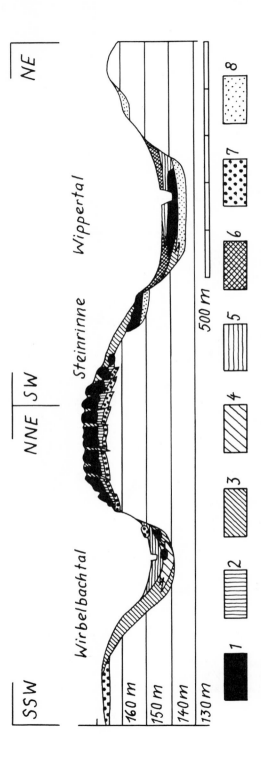

FIGURE 9.2
Geological cross-section through the quaternary depositions in the valley of Wipper, south of Bilzingsleben: 1, travertines; 2, basin loess; 3, younger loess; 4, solifluction gravel; 5, Holocene 'Aue' loam; 6, Holocene cliff gravel 7, terrace at 30 m distance to 'Aue'; 8, terraces at 20 m, 10 m, and 5 m distances to 'Aue' (Drenthe-, Warthe-, and Weichsel-terrace).

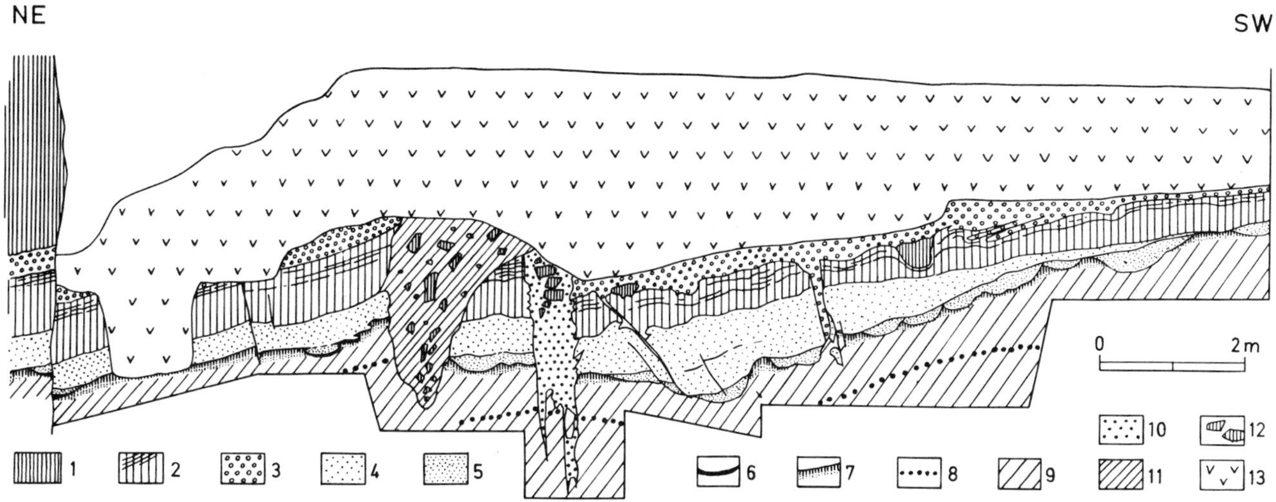

FIGURE 9.3

Cross-section through the travertine complex of Bilzingsleben with the fossiliferous layer in the area of excavation: 1, hard travertine; 2, lacustrine limestone; 3, loose travertine; 4, travertine sand in the delta channel; 5, travertine sand in the creek trough; 4 and 5, horizon of discovery; 6, bog-like formation; 7, clay-like soil; 8, limestone concretions; 9, basin loess; 10, interglacial crevice fill (travertine gravel); 11, glacial crevice fill (loess); 12, gravel pieces of travertine; 13, remains from the quarry.

138 *Homo erectus*

THE GEOLOGICAL-ARCHEOLOGICAL CIRCUMSTANCES

The travertine complex is constructed as follows (Figure 9.3). Lowest are glacial gravels of the river terrace. They transgress into a partly fluviatile, partly solid-fluid sedimentary loess of 3 m thickness. This loess is separated from the interglacial sediments above by a denudation-discordance. The sediments begin with a fine to medium-sized travertine sand of 60 cm thickness from which an approximately 50 cm thick stratum of light gray-to-white lacustrine limestone develops. A 3-to-4 m thick and 300-400 m large plate of hard-banked travertine ascends by incorporation of thin layers of loose and porous structured travertine. The plate is torn, distorted, and angled at its northwest-southeast directed crevices. The surface is deeply corroded by weathering and the fissures are filled with younger material, that is, with loess from the Saale/Riss and Weichsel/Würm glaciations, solifluction, and gravel. All dislocations and weatherings were caused after the sedimentation of the travertine complex during a glacial denudation. This not only caused the levelling of the slope and high plain of the former Middle Pleistocene river valley, but also led to a change in relief and in position of the travertine complex as a plateau. The plate of travertine acted as a hard, resistant cover.

It is the travertine sands at the base of the plate that are presently being excavated. Less rich strata are located in the overlying travertine where the occasional discoveries of the 1920s were made. The basal sands represent a broad, slightly sloping delta that had been built up by a creek running easterly from a nearby travertine water-well into a shallow lakebed. Some remains of the old creek bed are preserved below these sands.

The present excavation also includes the southern limits of the delta and its neighboring shoreline, which is covered by a fine travertine sand of 10 cm thickness. Whereas the discoveries in the delta and creek bed are of a parautochthonous nature and have been transported over a distance of several meters, the material from the shoreline can be considered partly autochthonous and may show the original details (Figure 9.4).

From this description the paleoecological situation can be deduced. The place was located on the shore of a small lake ideally suited for Middle Pleistocene man. The lake had been formed by travertine deposits at the foot of a valley below a rich spring. Hominids chose the shallow, sandy terrace at the mouth of the creek because it was free of vegetation as a result of the continually running water and because the location allowed easy access to the water itself. Dry areas of the delta served as places for repeated habitation during the Middle Pleistocene.

The archaeological material consists of a large number of stone artifacts, about 1500 kg of faunal skeletal remains from hunting activities, numerous tools of

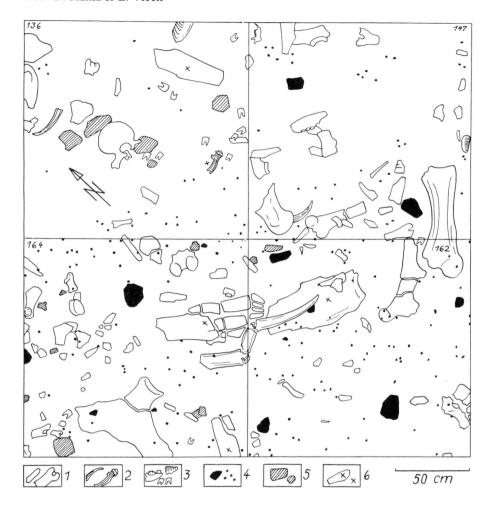

FIGURE 9.4
Distribution of the larger specimens on the shore area: 1, bones; 2, antler; 3, dental remains; 4, silex artifacts; 5, stone table surface; 6, bone or antler tool.

bone and antler, and hominid skeletal remains. Various fossil groups indicative of the geological and lithogenetic conditions, particularly valuable for paleoecological analysis, were recovered.

HOMINID REMAINS

Thus far, the following cranial and dental remains of fossil hominids have been discovered (dates in parentheses):

Bilzingsleben A1: large fragment of *os occipitale* (October 22, 1972; recognized April 17, 1974)
Bilzingsleben A2: small fragment of *os occipitale* which fits to A1 (August 6, 1974)
Bilzingsleben B1: fragment from middle region of *os frontale* (July 15, 1975)
Bilzingsleben B2: fragment from upper region of *os frontale* (August 25, 1976)
Bilzingsleben C1: right upper molar (July 3, 1976)
Bilzingsleben D1: fragment from *os parietale* (July 5, 1977)

The items were widely scattered over the excavated area of the site, encompassing 400 m^2 of the prehistoric surface of the delta (items A1, A2, B1, B2, C1) and shore (item D1). The following brief descriptions are based on an analysis by Vlcek (1978, 1979).

The Os Occipitale

This fragment of occipital bone (Figure 9.5), put together from pieces A1 and A2, is 115 mm wide and 75 mm high. It consists of the occipital plane (*planum occipitale*), the occipital torus (*torus occipitalis transversus*), and a large part of the nuchal plane (*planum nuchale*). The occipital plane forms a 40 mm high and 100 mm wide triangle. The attachment area for the occipital muscle (*m. occipitalis*) is separated from the mass of the occipital torus by a supratoral sulcus (*sulcus supratoralis*). This area of the occipital torus forms an elevation of 15 mm lateral to 19 mm median-sagittal. The lower border of the ridge is wavy and pulled up proximally in the median-sagittal plane. Inion and opisthocranion are located at the same point. The relatively flat nuchal plane and the low occipital plane form an angle of 108° at this point. Therefore, the occipital bone appears markedly bent.

Both halves of the nuchal plane show clear muscle markings. Below the occipital torus in the median-sagittal plane, the following muscle attachments are visible: *m. trapezius, m. semispinalis capitis, m. obliquus capitis superior, m. rectus capitis posterior major*. Another characteristic is the considerable thickness of the bone, which averages 9 mm.

Endocranially, the markedly drawn out occipital pole is clearly noticeable. The occipital part of the surface is bent in a characteristic way. On the occipital fossae the impressions of the gyri and cerebral juga are weakly visible; the surface above the cerebellar fossae (*fossae cerebellares*) is smooth. The sulci for

FIGURE 9.5
Bilzingsleben A1 and A2 occipital bone; occipital and left side views (.65:1).

the venous sinuses are strongly developed. The sagittal sulcus runs along the right of the median-sagittal plane; the left transverse sulcus is bent below the right.

The Os Frontale

Two pieces, B1 and B2 were found, which cannot be fitted together (see Figure 9.6 and 9.7). The larger of the two, B1, consists of the glabella (*pars glabellaris*), nasal part (*pars nasalis ossis frontalis*), and a small portion of the squama. It is 53 x 68 mm. The second piece, of 50 mm length and 30 mm width, can be oriented due to the preserved part of the coronal suture and the base of the frontal crest left of bregma.

The nasal part of the frontal bone is very large and features a huge nasal root. The shape of the nasofrontal suture is trapezoid (21 mm wide, 13 mm high). The most remarkable feature is the presence of a huge supraorbital torus, the surface of which is smooth. Although it is continuous across the glabella, there is a broad, smooth, and shallow depression above this part which blends with the surface of the frontal squama.

The size of the frontal ridge can be assessed from the following data: (1) 25 mm distance from nasion to frontal crest; (2) 28 mm from glabella to frontal crest; (3) 32 mm size of the torus at the lateral frontal incisure; (4) 21 mm height of the torus at nasion and at the medial frontal incisure.

142 Homo erectus

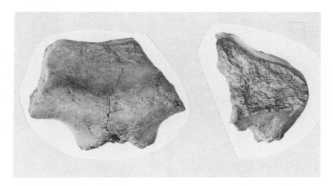

FIGURE 9.6
Bilzingsleben B1 frontal bone fragment; anterior and right side views (.65:1).

FIGURE 9.7
Bilzingsleben B2 frontal squama fragment; outer (left) and inner table views (.65:1).

The frontal squama has a smooth upper surface. It rises strongly and is 6-8 mm thick. The second fragment shows a simple zigzag configuration of the coronal suture; the latter is preserved for 30 mm.

The frontal sinuses are developed in a cabbage-like configuration, and their basal part is deeply inserted into the nasal part of the frontal bone. Radiographs show two large chambers which are separated not by walls but only by rib-like structures in the posterior and upper sides. The sinuses are 47 mm wide, 30 mm high, and 17 mm deep.

The inner cerebral face of the B1 fragment is separated into two halves, by an initially unclear then sharp and divided frontal crest. On both sides of the crest there are shallow impressions of the gyri and cerebral juga.

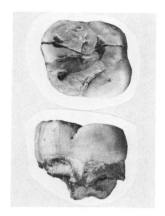

FIGURE 9.8
Bilzingsleben C1 molar (M^1 dx) (1.95:1).

The Right Upper Molar

Only the crown and portions of the neck are preserved; the roots are missing (Figure 9.8). The presence of mesial and distal contact facets indicate that the tooth is either a first or second molar. The crown is irregularly abraded. It is short and rhomboid in *Norma verticalis*. All four cusps are well developed but very much worn. The system of enamel folds is barely visible. In all three dimensions the pulp cavity is very wide and spacious.

The Os Parietale *Fragment*

This piece comes from the region above the 'asterion' of the right parietal bone. Its size is 55 x 47 mm. The 47 mm length is the lambdoidal suture which can be assembled with the corresponding part of the occipital bone A1 and A2.

PHYLOGENETIC STATUS

The Bilzingsleben cranial remains have been compared with other well-known remains of fossil man, particularly those that form a continuous chain in the evolution of *Homo* according to present acknowledged opinion. These include *Homo erectus* from Java, China, East and South Africa, the European finds from Swanscombe, Steinheim, Vértesszöllös, Arago-Tautavel, and Ehringsdorf, European Neanderthal Man, and finally the fossil and recent forms of *Homo sapiens*.

The most characteristic traits of the Bilzingsleben cranial fragments are as follows: a huge occipital torus; a markedly angled (108°) occipital bone with low occipital plane and flat nuchal plane; coincidence of inion and opisthocranion; a strong supraorbital torus with no interruption in the glabellar region; a broad and smooth glabellar depression; a generally broad and huge nasal root; a flat frontal squama; generally thick-walled cranial bones; a characteristically bent and drawn-out nuchal region of the endocranial cast.

On comparison, the Bilzingsleben occipital bone shows greatest similarity to the finds *Sinanthropus* III, *Pithecanthropus* VIII (Sangiran 17), and Olduvai Hominid 9. On the other hand, great differences exist when the bone is compared with the series of *Homo soloensis* and also Broken Hill as, for example, in the morphology of the occipital torus, in the thickness of the bone at inion, and in the considerable thickness of the occipital squama.

Compared with European finds, some similarity exists with the occipital bone from Vértesszöllös, but principal differences exist with regard to the Middle Pleistocene discoveries from Swanscombe and Steinheim as well as the Upper Pleistocene discovery from Ehringsdorf. These skulls do not show an occipital torus, and opisthocranion does not coincide with inion. The general contour of their occipital bones is curved rather than angled. On the whole they are *Homo sapiens*-like and principally distinct, as are the Neanderthal and remaining fossil forms of *Homo sapiens sapiens* when compared with the Bilzingsleben find.

The frontal bone from Bilzingsleben appears to be most similar to those of Olduvai Hominid 9 and *Pithecanthropus* VIII. It differs from the European forms from Steinheim, Arago, and Ehringsdorf, as well as from classic Neanderthal, in several morphological details of the glabella and supraorbital torus.

In the same way that the occipital and frontal bones can be grouped morphologically and metrically with the above mentioned *Homo erectus* finds, the molar from Bilzingsleben is morphologically within the variation of *Homo erectus*.

The comparison with *Sinanthropus* III and Olduvai Hominid 9 allow two variations for reconstruction of the skull from Bilzingsleben (Figure 9.9). Certain differences that exist with already known forms of *Homo erectus* justify recognition of the Bilzingsleben fossil as a geographical variant of *Homo erectus* in Europe and establishes it as a new taxon, *Homo erectus bilzingslebensis* Vlcek 1978. It proves the contemporaneity of two morphologically distinct forms of fossil men during the Middle Pleistocene in Europe: *Homo erectus* as represented by Bilzingsleben and Vértesszöllös, and a *Homo sapiens* form as represented in Swanscombe and Steinheim.

CULTURE OF *HOMO ERECTUS* FROM BILZINGSLEBEN

The cultural remains recovered from the strata have already been mentioned. Their facies is determined by their position on an open country site and by the

145 D. Mania & E. Vlček

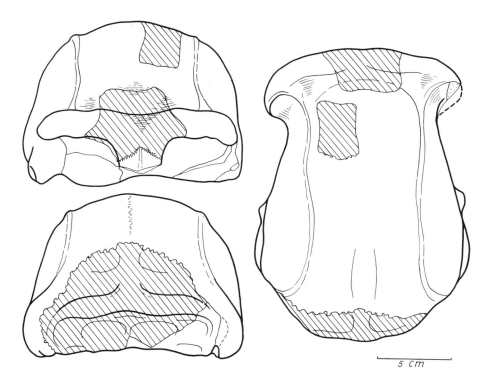

FIGURE 9.9
Reconstruction of the skull of the Bilzingsleben hominid, after the Olduvai Hominid 9 skull (Vlcek 1979).

limestone substrate of the embedding sediment. Noteworthy are thousands of artifacts made from hard cretaceous flint from the gravel of the Elster glaciation. The flint tools are relatively small, varying in length from 8 to 60 mm; two-thirds are broken cores and one-third are flakes. Mostly the edges, and rarely the surfaces, were worked (Figure 9.10). The fields of scars and splinters indicate that hammerstones from quartzite pebbles were used to produce the tools and for secondary working. The strike scars on the flakes, flake surfaces, and pieces with negative relief are mostly only known from the Clactonian technique.

The shape of the tools is heterogeneous. Rarely can one distinguish between different types, but the form of the retouched edges allows a categorization. These are straight, concave and convex retouched edges. Also, the edges are deep. A steep type of retouch predominates. Numerous edges are serrated, while

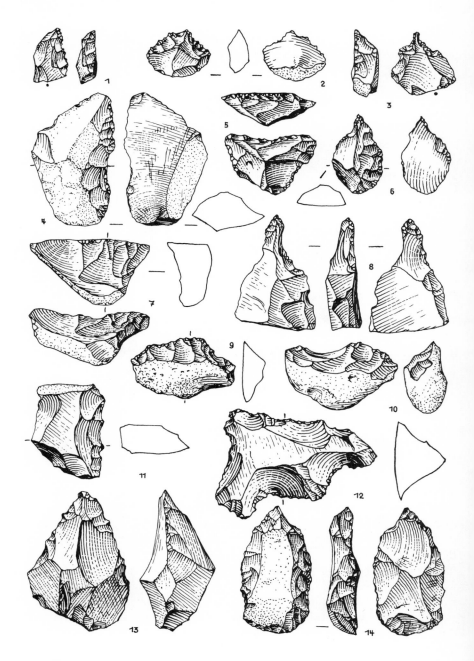

FIGURE 9.10
Flintstone tools from Bilzingsleben (after Mania 1977) (0.9:1).

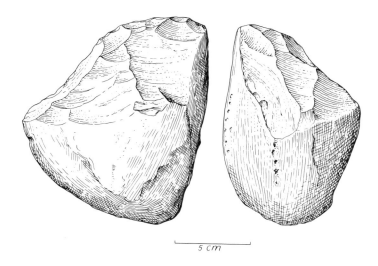

FIGURE 9.11
Quartzite pebble tool from Bilzingsleben.

others are scraper-like and often microlithic. Small tips that may be surface-retouched are reminiscent of the form on handaxes. Small triangular pieces with retouches on both sides exist. Some tools were produced from lemon-leaf-like flakes. Most of these tools probably served for cutting and wood shaving and for the working of other raw materials (but wood in particular).

There are also big and crude tools made from tough materials such as quartzite, mussel lime, and crystalline rock. They represent pebble tools (Figure 9.11) and include simple pebbles, one- or two-sided knapped pieces with hack-knife-like cutting edges or stump-cone tips, further primary big flakes, crude handaxe-like pieces or anvil blocks made of large pebbles, and plates with diverse traces of working.

Also found, surprisingly, was a series of bone and antler tools. Shed as well as taken antlers were manufactured into picks and clubs by breaking off different parts (Figure 9.12). Numerous holes and cracks on long bone shavings indicate intentional splitting of large extremity diaphyses done with the aid of small wedges and a hammer-like instrument. The shavings produced in this way were secondarily worked with a hammerstone like any flint tool (Figure 9.13). Recovered were straight and bowed scrapers, knife-like cutting instruments, chisels with splintered ends (up to 80 cm long), shaver-like tools, as well as needle and awl-like tools (Mania and Cubuk 1977) and pieces with ground-off and ground-on edges. All show traces of use such as rounded, polished surfaces,

148 *Homo erectus*

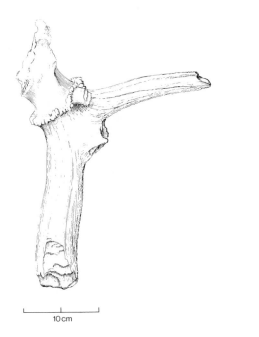
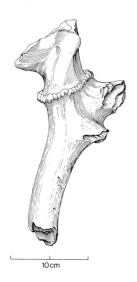

FIGURE 9.12
Antler chopping tools from Bilzingsleben.

parallel scrape or carve lines, cut lines, and splintering. In addition, there are flat anvil-like pieces with strike or pressure scars, or long cut marks (Figure 9.14).

The numerous pieces of stone refuse and the accumulation of stone, bone, and antler tools point to the fact that the tools were predominantly produced at the site and that work was done with these tools at different locations in the site.

Burned flint tools and some traces of charcoal at the site indicate the use of fire. Splinters of wood, structurally preserved by calcification, are probably refuse from wood tools or implements.

Also found at the site were the skeletal remains of hunted animals. These bones are mostly broken or crushed into small pieces. Prevalent are the remains of large mammals including forest elephants, the steppe and forest rhinoceros, bison, numerous deer, and large wild horses and wild oxen. Bear is frequently found among the remains. Other carnivores such as lion, wolf, panther, wild cat, etc. are rare.

Among smaller prey, beaver and the extinct beaver, *Trogontherium*, are common. Roe and wild pig are rare. Remains of large fishes, shells of large bird

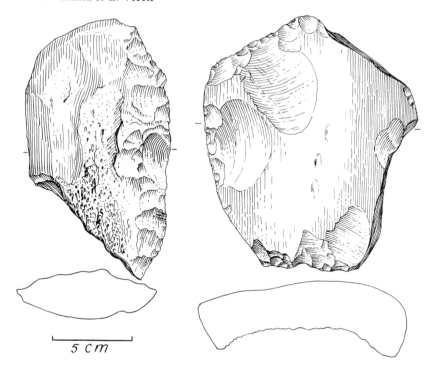

FIGURE 9.13
Bowed scraper (left) and chisel of bone from Bilzingsleben.

eggs, and pieces of bivalves occur. The latter two items suggest food-gathering activities. An analysis of the faunal remains in terms of hunting economy has still to be undertaken.

PALEOECOLOGY

The paleoecological conditions at the Bilzingsleben site can be assessed not only from the lithogenetic and geomorphological evidence but also from the preserved fossil forms. Of importance in the latter group are the impressions of plants in the travertine, the vertebrate faunal remains, and the mollusks and ostracods. Small mammals and impressions of insects are rare.

The geographic position of the site in the Middle Pleistocene valley of the Wipper River has already been briefly described. The overall conditions of vegetation at the time of the travertine buildup can be deduced from the plant remains. Represented are members of the thermophile deciduous mixed forests,

150 *Homo erectus*

FIGURE 9.14
Base-plate with cut marks, from the pelvic bone of a rhinoceros, Bilzingsleben.

such as *Quercus, Acer, Alnus,* and *Tilia,* and numerous shrubs, i.e. *Corylus, Cornus,* but also *Pyracantha* and *Buxus.* These latter species are not presently found among the natural flora of northern middle Europe but occur in the Mediterranean region. They suggest a warmer climate, on the average, than is presently experienced in the Bilzingsleben region. Similarly, the malacological analysis points to a previously warmer climate. Among the normally middle-European mollusks at the site there are some exotic ones which today are found in southeastern Europe and the Mediterranean region (*Aegopis verticillus, Discus perspectivus, Theoduxus serratiliniformis*). The distribution of the molluscan fauna from the travertine complex agrees well with that from other Holsteinian interglacial occurrences of the Elbe-Saale region (Mania and Mai 1969; Mania 1973).

The sediment facies and the floral and mollusk communities indicate not only the general interglacial climatic conditions but also suggest that the surroundings of the site were not completely covered by trees during the Middle Pleistocene. Forests must have been interspersed with open areas, a landscape that offered an optimal environment for large herbivores and, consequently, successful hunting conditions for primitive man.

SUMMARY

Hominid cranial remains that can be classified as *Homo erectus* have been described from a Middle Pleistocene travertine complex in northern Thuringia at Bilzingsleben, German Democratic Republic. The site suggests the repeated presence of Middle Pleistocene hominids near the mouth of a creek on the shore of a lake. Stone artifacts belonging to a small-tool, flint-flake, and chip industry are present. Crude pebble tools made of quartz, quartzite, limestone, and crystalline are also present. Numerous specialized tools consist of antler and intentionally split bones. The manufacturing technique on the bone and stone tools is the same. Among the skeletal remains of animals are those of the forest elephant and rhinoceros, the European deer, bear, beaver, and bison. Remains of *Trogontherium* are noteworthy. Plant remains indicate the presence of thermophile mixed forest and shrubs of middle European origin, and there is a strong Mediterranean influence. Molluscs, pointing to strong relationships with the Mediterranean and southeast European region, are also present. Climatic conditions during the time of *Homo erectus* in northern middle Europe were warmer and milder than today.

10

F. CLARK HOWELL

Some views of *Homo erectus* with special reference to its occurrence in Europe

Cet article repose sur le texte révisé de la conférence prononcée par l'auteur au symposium. Son exposé portait sur les problèmes que pose l'étude de l'*Homo erectus* en Europe. En ce qui concerne la nomenclature, Howell se dit d'accord avec de Lumley pour qualifier ces premiers hominiens européens de prénéanderthaliens. Mettant en lumière certains spécimens de fossiles, il étudie leur importance dans le cadre phylogénétique de l'homme. En conclusion, il attire notre attention sur certains domaines qui nécessiteront une investigation plus approfondie si nous voulons connaître la place et le rôle des premiers hominiens européens dans le processus général de l'évolution de l'homme.

The species *Homo erectus* is well defined temporally and morphologically. The time range is reasonably well established, not necessarily in its first major known source area, Asia, but certainly in Africa, particularly in East Africa. It appears to be a species that had a duration roughly on the order of 750,000 to 800,000 years, from as early as about 1.4 million to 1.5 million years ago, judging from the evidence from East Turkana. On the other hand, I think that our surmises about ages of sites on the mainland of Asia may be as much as 100-500 percent in error. There is even some direct evidence in Asia to suggest now that our 'guesses,' which were made about twenty years ago, were widely in error.

Morphologically, the best postcranial evidence is, again, not from Asia, although there is a certain amount of evidence from Chou Kou Tien which suggested a long time ago that the postcranial skeleton was not precisely like that of *Homo sapiens*. Subsequently, the recovery of material, particularly from East Africa — at Olduvai Gorge and more recently at East Turkana — has shown that it is a skeletal structure that is *Homo*-like but not *Homo sapiens*-like. The cranial, dental, and facial morphology is well established and it is very distinctive. There are a set of characteristics, obviously many of them interrelated, that define the species in my mind clearly. These characteristics are extremely important in characterizing the species. If we use such criteria to define the species, then the evidence is that *Homo erectus* does not exist in Europe. That doesn't mean that it didn't, but that we haven't found it there at present.

Take, for example, the mandible from Montmaurin in France in the Haute Garonne. Before continuing, I would caution that unless you already have well-established morphological criteria on the basis of associated skeletal remains, please do not compare mandibles without skulls to skulls without mandibles. Don't compare feet without hands to hands without feet. We do not know how these parts interrelate and too often there is a wide-ranging extrapolation as to how these characters should or shouldn't work when we really don't know. Australopithecines have taught us that lesson. *Homo erectus* has taught us that lesson in a different way. Montmaurin, when it was described, was shown clearly to present a series of characteristics that are morphologically distinct from Neanderthal and to show characteristics that are 'archaic.' It shows a series of characteristics in the robustness of the mandibular body and the structure of the symphysis and the ascending ramus and, in part, in the molar dentition that diverges from the Neanderthal, and so the use of the term 'Anteneanderthal' is extremely acceptable to me. It is Anteneanderthalian in terms of time and Anteneanderthalian in terms of morphology. It also diverges, in ways that have already been published, from known ranges of variation of *Homo erectus* samples from Asia and samples from Africa.

The specimens from Arago that Dr M.-A. de Lumley has mentioned are among the most important discoveries ever made in europe and they are of great interest. Again, the remarks about morphology are comparable to those one might make about montmaurin, except in this case we fortunately have portions of the cranial vault including the anterior part and the face. the mandible shows the kind of features that you see in Montmaurin. They also show a number of resemblances to Mauer. But in a number of morphological details such as in the ascending ramus, in the symphysis, and in the dentition itself there are divergences from *Homo erectus*. The facial skeleton shows divergences from *Homo erectus* morphology, for example, in the nasal region and the shape and form of the orbit and the supraorbital region. The upper part of the face in relations to the frontal bone is unlike that found in known *Homo erectus* samples. In my view, we have here a total set of interrelated features that diverge significantly and occur as a whole set of features slowly emerging in populations. In this case we're talking about populations in the time range of several hundreds of thousands of years ago that were on the brink of the penultimate glaciation or the latter part of the so-called Great interglacial.

Swanscombe diverges significantly in its morphology from all known *Homo erectus* individuals. Steinheim shows a whole set of characters all of which approach those of the Swanscombe morphology. There is no question that Bilzingsleben, dated at 340,000 BP [or about 230,000 BP on newer ^{230}Th^{234}U dating of associated traveltine], as determined by preliminary estimates from amino acid racemization dating, is between advances of the antepenultimate and pen-ultimate glaciation. It is in the great interglacial, on the basis of stratigraphy found in some of the type areas of the great interglacial, the main ice sheet in Germany. There is an extensive fauna that provides a very elaborate body of data all of which give a broadly Great interglacial age. I do not know any Pleistocene geologist who believes that the Great interglacial is as old as 340–350 thousand years, so something is awry here. There are far more advances and retreats and far more interglacial major periods of warming, more or less in the northern hemisphere and especially in Europe where there have been less studies, than anybody ever suspected. So one should be extremely careful about how one puts these occurrences into some sort of framework. The best dated occurrences in my opinion, and I agree completely in this respect with Marie de Lumley, are the major occurrences she has touched upon in France. There are extremely good and consistent data for most of those localities, particularly those that range through the Riss glaciation.

Regarding Vértesszöllös, Professor Thoma has clearly demonstrated beyond any reasonable doubt that this is not a *Homo erectus* individual in the normal sense of that term. Vértesszöllös is older than anything we have seen thus far. It is clearly older than Bilzingsleben on the basis of its position in an extremely well-defined eastern European transcarpathian faunal succession. There is no question that it occupies some position in the major interglacial and partial glacial phase prior to the great interglacial.

Mauer is within the Brunhes normal epic. It is less than 700,000 years old, or if you prefer, less that 690,000 years, it is extremely well defined in age based on a very fine faunal succession that includes microvertebrates in the Rhine-Main Basin. Many of us have argued for years that it diverges in a number of characteristics from the normal, expected, and now well-established range of variation in *Homo erectus* populations.

Petralona is a cave just outside of Thessalonica in northern Greece. The Petralona skull found there was originally published to be a Greek Neanderthal, the first one really known. A beautifully preserved specimen, it has never been fully and properly studied, and it has never been completely cleaned. We are in the first one really known. In recent years the specimen has been largely, and beautifully, cleaned by Professor J. Melentis.

The cave site of Petralona needs much further investigation. The faunal picture is complex. In Greece, fauna of this kind are not well known. However one can go slightly to the north into Hungary and adjacent areas, or to Czechoslovakia, or to the west and south to Yugoslavia where more data are known, and make comparisons. There is no question that the Petralona fauna is pre-Upper Pleistocene. The late Otto Sickenberg studied the fauna twice. He first attributed the fauna as belonging to a last interglacial age. Later he changed his interpretation of it to an age as early as the Biharian, a range of time that precedes the great interglacial, and is 'equivalent' to the Mindel complex. Sickenberg's work demonstrated very clearly, particularly with deer, equids, carnivores, and possibly rodents that this kind of fauna had a Biharian aspect in the broadest sense. That means that the age of the fauna could be on the order of the age of Vértesszöllös, or perhaps younger.

The Petralona skull itself was recognized by other people to diverge from Neanderthal. It should be considered Anteneanderthal in the sense of Marie de Lumley's definition. Others consider it to be a *Homo erectus*. Morphologically it diverges significantly from *Homo erectus* in all those features that distinguish the cranial-facial morphology. Moreover, the occipital region lacks all distinguishing features found in all *H. erectus* specimens. There is a fair amount of frontal constriction but not to the extent or of the nature found in *Homo erectus* populations. The skull lacks the parasagittal thickening and the sagittal torus that is found in practically all *H. erectus* populations. Helmuth Hemmer in Mainz did some comparisons and stated that cranial-facial proportions of the Petralona skull were not Neanderthal, but rather diverged in the *Homo erectus* direction. However it does not quite follow that because of a few combinations of proportions the skull can be classified as *Homo erectus*. I think Dr Hemmer would agree with that. In my view, there are remarkable similarities between the Petralona and the Arago specimens, with respect to mid-facial region, the nasal bridge, the shape of the orbits, and the structure of the supraorbital region and its relation to the frontal. In all these respects the Petralona skull shows significant divergences from the pattern we know of as *Homo erectus*.

How are we to understand the origin of *Homo* populations in Europe? Are they in fact an emergent set of populations that arose from a *Homo erectus* base? Are we dealing with populations that are merely geographic variants, not necessarily temporal, or more characteristic *Homo erectus* populations in Asia and Africa? Is it that *Homo erectus* as we know the taxon from present evidence never reached Europe in that state? Had the first human populations into Europe already passed the '*Homo erectus*' level or grade of organization? It is extremely important to understand the nature of the morphology and variations in morphological patterns. In addition it is essential that we find postcranial remains in order to understand the locomotor structure and postcranial proportions and details of the skeleton in these populations (part of an ilium is known from Bilzingsleben). My prediction is that they will not look like *Homo erectus* postcranial parts.

In summary, we do not understand the nature of the Anteneanderthal of Europe, either archaeologically or otherwise for that block of time that precedes the great interglacial. We have only glimpses of it. We do not really know when hominids entered Europe for the first time or where they came from. And we still do not know what these European early hominids looked like. If it was on the order of a million years ago, presumably they were at the *Homo erectus* grade. This is based on inference, because that's the grade that we know them in Africa where we have good dates, and also in Indonesia where the age estimates are reasonable and consistent with what is known both stratigraphically and from radiometry. Prior to that range of time I am less happy than are some of my colleagues about the presence of hominids at all on the European continent.

Here, then, are a set of problems in human evolutionary studies that one could not have talked about fifteen or twenty years ago because of the lack, or scantiness, of the evidence. We have an increasing body of evidence that deserves to be looked at in several different ways and compared with the increasing body of evidence from other parts of the Old World. This increasingly important evidence, and the emerging set of problems based on it, should be studied and interpreted in its own right rather than merely pigeonholed into previously established traditional categories.

11

J.-J. JAEGER

Les hommes fossiles du Pléistocène moyen du Maghreb dans leur cadre géologique, chronologique, et paléoécologique

The geology and relative chronology of the Maghreb Middle Pleistocene hominid sites are discussed in light of recent data. Correlations are proposed between the stratigraphic framework of Atlantic Morocco and a marine oceanic Pleistocene framework that permit the attribution of absolute ages to several Maghreb sites. The paleoenvironment is broadly outlined. Two distinct hominid populations, which need not have been directly linked phyletically, are identified. One is the Ternifine population, which is older and which is comparable to other *Homo erectus* populations. The other is the Abderrahman-Rabat-Salé group, which possesses several unexpected modern characteristics in addition to a great number of *H. erectus* characteristics that justify its inclusion in that taxon.

The phyletic relations between the latter group and the modern-man fossils of the Maghreb are discussed. The recent age of the Sidi Abderrahman-Rabat-Salé group (220,000 to 150,000 years ago) and the persistence of numerous primitive characteristics of the *H. erectus* type suggest that a progressive evolution of the group into *Homo sapiens* is unlikely since it would have had to occur in a very short time. The modern characteristics of the group are, therefore, interpreted as having resulted from an unsuccessful attempt at sapientization. The evidence the fossils present, relative to the modalities of such an evolution, is of considerable interest for the study of this phenomenon.

L'Afrique nord-occidentale, ou Maghreb, située au coeur de l'ancien monde, constitue un carrefour biogéographique entre l'Afrique et l'Europe d'une part, l'Afrique et l'Asie d'autre part. Cette situation privilégiée confère à l'étude de l'évolution de l'homme dans cette région un intérêt tout à fait exceptionnel.

Les plus anciens hominidés connus de cette partie du globe proviennent tous de dépôts d'âge Pléistocène moyen et peuvent être attribués à *Homo erectus*. Bien que les hominidés plus anciens n'y aient pas encore été découverts, leur présence est attestée par de nombreuses découvertes de restes lithiques de type primitif.

LE CADRE CHRONOLOGIQUE

LES DÉPÔTS
Les hommes fossiles du Pléistocène moyen du Maghreb proviennent de deux types de dépôts bien caractérisés. Le gisement de Ternifine (Arambourg et Hoffstetter 1954, 1955, 1963; Arambourg 1956, 1957, 1963) correspond à un dépôt lacustre alimenté par une résurgence de type artésien située au milieu d'une plaine alluviale. Les dépôts sont constitués d'argiles lacustres à la base, surmontées par des sables remaniés du Miocène terminal sous-jacent. Tous les restes humains et la plupart des mammifères fossiles récoltés proviennent des argiles lacustres de la base. Les niveaux sableux sus-jacents ont cependant également livré quelques restes lithiques et osseux.

Tous les autres gisements à Hominidés sont situés sur le littoral atlantique marocain. Ils proviennent soit d'éolianites (Rabat, Salé), soit de remplissages de grottes creusées par la mer dans ces éolianites et comblées ou par des dépôts marins ou par des argiles rouges de décalcification, restes d'anciens sols quelquefois interstratifiés entre les différents niveaux d'éolianites. Seuls les restes de la carrière Thomas III proviennent d'une cavité de dissolution dans les éolianites, non remplie ultérieurement.

Il apparaît donc bien clairement que, si Ternifine pourrait correspondre à un véritable sol d'habitat dont les structures n'ont pas été relevées au cours des fouilles, tous les autres sites à restes humains du rivage atlantique marocain ne peuvent être assimilés à de tels sols. Il s'agit pour ces derniers de débris entraînés par des prédateurs dans les dunes ou dans les grottes du littoral.

CHRONOLOGIE RELATIVE

a *Données Stratigraphiques*
La chronologie relative du Pléistocène du Maghreb a été définie à partir des cycles sédimentaires du littoral atlantique marocain. Les nombreux travaux qui s'y réfèrent sont résumés dans la Figure 11.1.

Les sites du littoral marocain peuvent être replacés dans ce cadre chronologique. Le gisement de Sidi Abderrhamane (Biberson 1956, Arambourg et Biberson 1955, 1956) provient d'une grotte creusée dans les éolianites de l'Amirien et comblée par des dépôts marins de l'Anfatien G_0, G_1 et G_2; c'est dans les premiers niveaux continentaux superposés à la lumachelle G_2 qu'ont été découverts les restes humains. Pour Biberson (1961) ce niveau continental pourrait encore être contemporain du maximum de la transgression anfatienne G_2. Néanmoins, compte tenu des définitions théoriques proposées en 1971 par Biberson, ce dépôt est assigné au Tensiftien inférieur.

161 J.-J. Jaeger

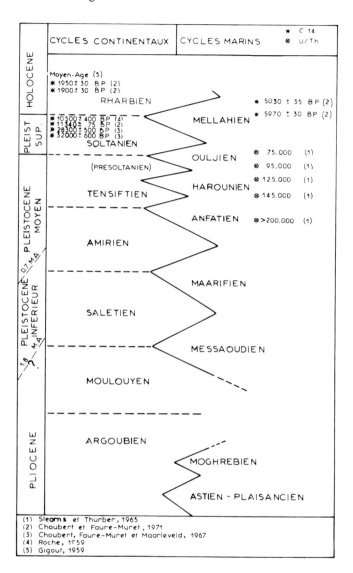

FIGURE 11.1
Cadre stratigraphique classique du Pléistocène du Maghreb.

Le gisement de la carrière Thomas I, qui a livré une mandibule (Ennouchi 1969c) étudiée en détail par Sausse (1975), provient d'une grotte (Sausse 1975; Jaeger 1975a, 1975b) creusée dans les éolianites de l'Amirien moyen. La mer anfatienne, à son maximum transgressif G_2, pénètre dans cette carrière en formant un minuscule golfe et vient buter sur la base de la grotte sans qu'il soit possible de vérifier si le creusement puis le remplissage de cette dernière sont antérieurs, contemporains ou postérieurs à l'épisode transgressif G_2 de l'Anfatien.

La grotte est néanmoins colmatée par des éolianites du Harounien-Rabatien qui permettent de situer géologiquement ce remplissage entre l'Amirien supérieur et le Tensiftien moyen. Nous verrons ultérieurement que la faune de micromammifères associée permet de préciser cette datation.

Le gisement de la carrière Thomas III (Ennouchi 1972, 1976) correspond à une petite cavité de dissolution creusée dans les éolianites de l'Amirien moyen, non comblée par des sédiments et recouverte par la transgression anfatienne.

La géologie du littoral des environs de Rabat est moins bien connue que celle de Casablanca, 4 cycles marins sont visibles: l'Ouljien, le Harounien-Rabatien, l'Anfatien G_2, et un cycle inférieur dont la nature exacte n'a encore pu être précisée.

La transgression anfatienne qui correspond, vraisemblablement uniquement au niveau G_2, bien que ce dernier point ne soit pas encore explicitement démontré, constitue l'essentiel des dépôts marins du Quarternaire des environs de Rabat. Elle atteint dans cette région une altitude maximale de 30 à 40 m. Le crâne de Salé provient du milieu d'une importante éolianite qui s'est développée en arrière de ce maximum transgressif (Jaeger 1975c). Il est donc vraisemblablement contemporain des restes humains de Sidi Abderrahman, c'est-à-dire qu'on peut lui attribuer un âge Tensiftien inférieur, en fait sensiblement contemporain du maximum transgressif de l'Anfatien G_2.

La Figure 11.2 illustre la position stratigraphique de l'homme de Salé. L'homme de Rabat (Marçais 1934), dont la position stratigraphique a été considérablement discutée, provient également d'une éolianite plus récente que cette correspondant au crâne de Salé, occupant la même situation stratigraphique que l'éolianite n° 4 de la Figure 11.3. Il s'agit donc très vraisemblablement d'un dépôt tensiftien moyen. Toutefois, l'hypothèse de le rattacher aux éolianites du Harounien-Rabatien régressif ne peut encore être définitivement écartée. Cette dernière interprétation rajeunirait encore légèrement l'âge de ce fossile.

Les restes humains de Ternifine ne peuvent être attribués directement à l'un ou l'autre des cycles marins ou continentaux du Maroc atlantique. De ce fait, les

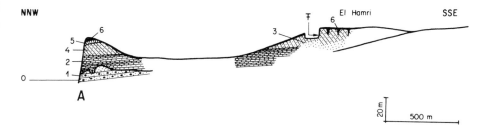

FIGURE 11.2
Coupe géologique du littoral au niveau du gisement de Salé (Maroc).
F: emplacement du crâne humaine. 1: niveau marin d'âge indéterminé raviné par l Anfatien.
2: Anfatien. 3: Dune correspondant au maximum transgressif de l'Anfatien. 4: Dune de l'Anfatien régressif. 5: Croûte calcaire. 6: Limons rouges soltaniens.

corrélations qui peuvent être établies ne peuvent reposer que sur les vertébrés ou l'industrie lithique.

b *Données Paléontologiques*

Grands mammifères. Les grands mammifères du Pléistocène moyen du Maghreb n'ont pas fait l'objet de travaux approfondis.

Le Tableau 11.1 indique pour la plupart des sites à hominidés et autres sites archéologiques, la répartition des différentes espèces.

En ce qui concerne les données biostratigraphiques, celles-ci sont restreintes à la distinction de deux faunes différentes, l'une ancienne bien représentée à Ternifine, l'autre récente, de la fin du Pléistocène moyen. Durant cette période, l'histoire de ces faunes est essentiellement marquée par l'immigration de formes eurasiatiques qui sont respectivement *Cervus* pour l'Anfatien et *Sus scrofa* pour le Présoltanien (Jaeger 1975b). On note également une augmentation progressive du nombre de taxons caractéristiques de savanes sèches comme les Gazelles et le Phacochère dont il sera fait état ultérieurement. Enfin Arambourg (1938) a décrit une lignée évolutive qui conduit d'*Elephas atlanticus atlanticus* à *E. atlanticus maroccanus*. Cette évolution ne permet toutefois pas d'établir des subdivisions plus fines que celles proposées ici.

Industrie lithique. Biberson en 1961 a reconnu dans l'Acheuléen des formations littorales atlantiques du Maroc une série de stades numérotés de I à VIII. Ces stades définis en provenance de différents niveaux de l'Amirien inférieur pour le stade I, au Présoltanien pour le stade VIII, ne peuvent toutefois être

164 *Homo erectus*

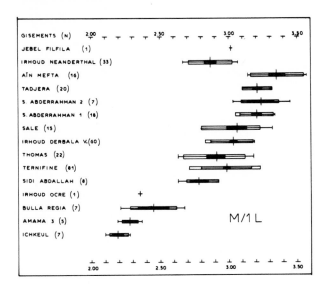

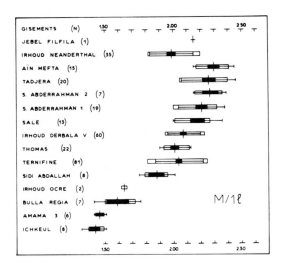

FIGURE 11.3
Dimensions des M1 supérieures et inférieures des différentes populations du Muridé *Paraethomys* au cours du Pléistocène.

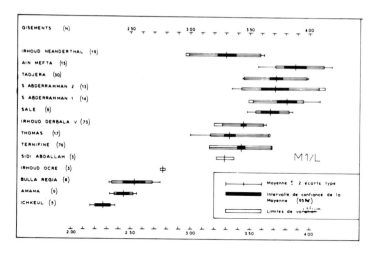

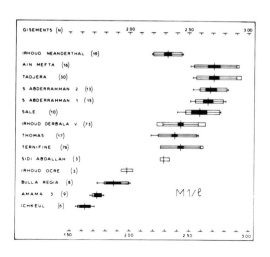

166 Homo erectus

TABLE 11.1
Répartition stratigraphique des faunes de grands mammifères du Pléistocène moyen du Maghreb

	Ainhanech	Ternifine	S.T.I.C.	Grotte des Ours et al.	G1	G2	Do	D1	Grès de Salé	Grès de Rabat	D2 Presoltanien
			Inf. Amirien Sup.		Anfatien		Tensiftien				
Elephas moghrebiensis	•										
Elephas iolensis			•	•	•						
Elephas atlanticus		•								•	
Ceratotherium simum	•	•	•	•	•	•	•	•	•	•	•
Equus mauritanicus		•	•	•	•	•	•	•	•	•	•
Equus (Asinus) tabeti	•										
Stylohipparion lybicus	•										
Omochoerus phacochoeroïdes	•										
Afrochoerus sp.		•									
Phacochoerus aethiopicus										•	
Sus scrofa											•
Hippopotamus amphibius	•	•	•		•	•					
Gazella sitifensis	•										
Gazella sp.		•									
Gazella atlantica				•		•	•			•	•
Gazella dorcas							•				•
Gazella cuvieri							•				
Bos bubaloïdes	•										
Bos praefricanus	•										
Bos primigenius				•		•				•	•
Oryx el eulmensis	•										
Oryx sp.		•									
Hippotragus sp.				•		•					
Redunca maupasi		u	•					•			•
Alcelaphus probubalis			•	•	•	•	•	•		•	•

TABLE 11.1 continued

	Ainhanech	Ternifine	S.T.I.C.	Grotte des Ours et al.	G1	G2	Do	D1	Grès de Sale	Grès de Rabat	D2 Presoltanien
			Inf. Amirien Sup.			Anfatien		Tensiftien			
Alcelaphus sp.	•	•									
Connochaetes taurinus			•					•	•	•	•
Connochaetes sp.		•									
Taurotragus sp.		•									
Taurotragus debianus			•								
Gorgon mediterraneus	•										
Numidocapra crassicornis	•										
Rabaticeras arambourgi										•	
Cervus sp.					•	•					
Libytherium maurusium	•										
Giraffa pomeli	•										
Giraffa sp.		•									
Camelus thomasi		•									
Crocuta crocuta	•	•	•				•			•	•
Hyaena hyaena		•			•		•	•			•
Canis aureus		•			•		•				•
Canis cf. atrox	•	•									
Mellivora sp.		•									
Mellivora cf capensis							•				
Ursus arctos		•		•	•						
Panthera leo		•									
Panthera pardus		•									
Homotherium cf. latidens		•									
Cercopithéciné indet.		•									
cf. Simopithecus		•									
Homo erectus		•					•		•	•	

utilisés comme indicateurs biostratigraphiques ne s'agissant pas de matériel soumis aux lois biologiques de l'évolution.

Cette succession peut de ce fait être sans aucun rapport avec les données stratigraphiques et peut avoir été déterminée considérablement par le hasard de l'échantillonnage (Freeman 1975).

Micromammifères. L'étude de l'évolution des Rongeurs du Pléistocène du Maghreb a permis d'établir une échelle biochronologique continentale beaucoup plus précise qu'aucune des autres échelles préexistantes. Elle a été établie grâce à l'évolution de deux genres de Rongeurs, un Muridé, *Paraethomys* et un Arvicolinae *Ellobius*.

Le genre *Paraethomys* définit un Muridé qui apparait au Miocène supérieur de part et d'autre de la Méditerranée. Alors qu'en Europe sud-occidentale il évolue jusqu'à la fin du Pliocène, sans jamais représenter une part importante de la biomasse des Rongeurs, il connaît en Afrique du Nord, du Pliocène au Pléistocène supérieur, un développement considérable qui conduit à en faire le Rongeur le plus commun et le plus abondant du Pléistocène du Maghreb. Son évolution est marquée par une adaptation progressive à une alimentation strictement herbacée, à partir d'un ancêtre à régime vraisemblablement omnivore. Les transformations morphologiques des couronnes dentaires et de l'appareil masticateur ont permis à Jaeger (1975a) de reconstituer en détail la phylogénie de ce genre durant le Pléistocène moyen. Deux espèces sont représentées pendant cette époque, *P. tighennifae* à Ternifine d'une part, *P. darelbeidae* dans les gisements présoltaniens d'autre part. Cette dernière espèce dérive, directement ou indirectement, de *P. tighennifae*.

La population de Thomas I, par ses dimensions et par ses caractères morphologiques se rattache indiscutablement à *P. tighennifae* et diffère significativement de *P. darelbeidae* du Présoltanien, plus grand et plus évolué pour certains caractères (Figure 11.3 et 11.4).

Ellobius est un rongeur fouisseur encore représenté actuellement en Asie. Les plus anciens *Ellobius* du Maghreb proviennent du gisement de Ternifine. La population de cette localité présente de nombreux caractères primitifs. Ce genre persiste jusqu'au Présoltanien où il est représenté par une forme de plus grande taille, possédant de nombreux caractères morphologiques plus évolués par rapport à celle de Ternifine.

La population de Thomas I est en tous points intermédiaire entre celle de Ternifine et celle du Présoltanien (Figure 11.5). Compte tenu du fait que, d'après les données géologiques, l'âge du remplissage de Thomas I est postérieur à l'Amirien moyen et antérieur à l'Harounien-Rabatien, les micromammifères permettent de proposer un âge un peu plus récent que Ternifine et nettement

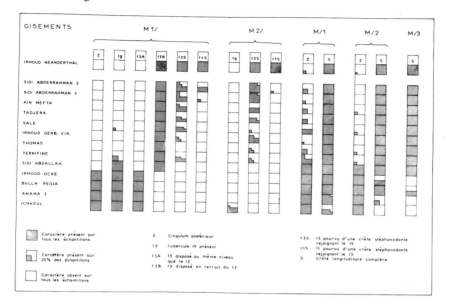

FIGURE 11.4
Modifications de fréquence des caractères chez les différentes populations du Muridé *Paraethomys* au cours du Pléistocène.

plus ancien que le Présoltanien. Cet ensemble de données converge pour attribuer un âge Amirien supérieur au gisement Thomas I.

Néanmoins, il n'est pas encore possible de confirmer l'âge Amirien inférieur attribué par Biberson, sur la base de l'industrie lithique, au gisement de Ternifine. Les faunes de grands mammifères ont une composition distincte, en raison des différences écologiques entre le littoral atlantique et les hauts plateaux algériens. Toutefois Arambourg considère que la présence de *Bos primigenius* à Casablanca et l'absence de formes relictes du Pléistocène inférieur (*Homotherium, Afrochoerus, Simopithecus*, etc.) attestent un âge plus ancien pour Ternifine. Cette dernière hypothèse n'est pas à exclure a priori et nous paraît reposer sur des arguments plus valables que ceux de Biberson.

L'absence de microfaune dans les dépôts de l'Amirien inférieur de Casablanca ne permet pas de choisir entre ces deux interprétations.

CHRONOLOGIE ABSOLUE ET CORRÉLATIONS

Les données de chronologie absolue relatives au Pléistocène moyen du Maghreb sont extrêmement réduites. Il n'y a aucune donnée relative au paléomagnétisme

170 *Homo erectus*

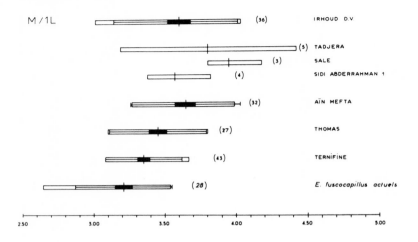

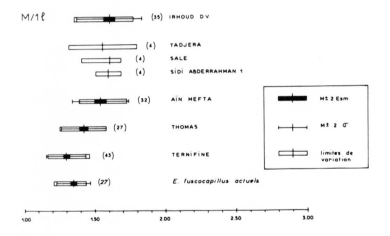

FIGURE 11.5
Dimensions des M1 supérieures et inférieures chez les différentes populations d'*Ellobius* du Pléistocène moyen du Maghreb.

M1/L

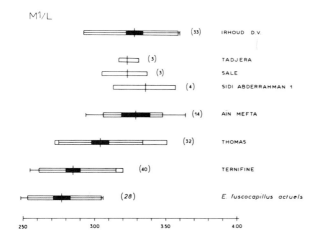

M1/ℓ

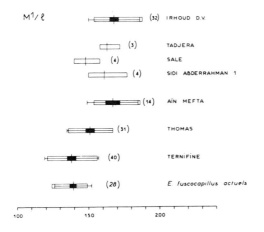

malgré l'intérêt qu'il y aurait de pouvoir situer dans la série atlantique du Maroc, la limite Brunhes–Matuyama. Les seules datations absolues sont relatives à la méthode Uranium-Thorium (U/Th). Elles ont été établies par Stearns et Thurber en 1965 (Figure 1.1) à partir de l'analyse de coquilles de gastéropodes marins de l'Ouljien, du Harounien-Rabatien et de l'Anfatien. Elles permettent d'attribuer un âge de 75,000 à 90,000 ans pour l'Ouljien, de 125,000 à 145,000 ans pour le Harounien-Rabatien et de plus de 200,000 ans pour l'Anfatien G_1 et G_2. Ces âges, comme tous ceux qui ont été obtenus grâce aux Gastéropodes, ont été critiqués et remis en cause par Kaufman et al. (1971) qui ont émis les plus grandes réserves à leur égard. Toutefois la constance de certains résultats et leur provenance de niveaux bioclastiques indurés analogues aux coraux ont conduit Butzer (1975) à minimiser ces critiques.

Le gisement de Ubeidiya au Proche-Orient représente un autre point de repère. En effet, sa faune de Rongeurs est plus primitive que celle de Ternifine (Jaeger 1975a) de même que son industrie lithique; la microfaune d'Ubeidiya a été rapportée par Janossy (1975) à la phase faunique de Betfia qui peut être, pour ses niveaux les plus anciens, corrélée avec l'Eburonien des Pays-Bas (Van der Meulen et Zagwijn 1974) lui-même plus récent que 1.6 MA indication apportée par le paléomagnétisme (Van Montfrans 1971). En outre, un équivalent latéral de la partie supérieure de la formation a été daté radiochronologiquement de 0.640 ± 0.12 MA (Siedner et Horowitz 1974). De ce fait, l'âge du site fossilifère est vraisemblablement compris entre 0.64 et 1.6 MA. Une corrélation peut être également établie avec les échelles du Pléistocène moyen des fonds océaniques (Figure 11.6). L'étude des modifications isotopiques des carbonates des eaux marines ont permis d'établir dans des séries océaniques continues l'histoire bioclimatique du globe. Ces échelles sont actuellement très comparables entre elles, quelle que soit leur origine géographique (Duplessy et al. 1976), et sont relativement bien calibrées. Compte tenu des datations U/Th obtenues pour le littoral marocain, il est relativement aisé de mettre en évidence la correspondance de l'Ouljien et de l'Harounien avec la période 5 et de l'Anfatien avec les périodes plus anciennes. En outre, le fort refroidissement mis en évidence, dans l'Anfatien G_1 par la prolifération du gastéropode *Littorina littorea* associé à la disparition des espèces d'eau chaude ou tempérée (Biberson 1961; Brébion 1976) nous permet de corréler cet épisode au stade 10 de l'échelle isotopique prise comme référence dans le présent travail (Shackleton et Opdyke 1973). Ce stade 10 est considéré par Duplessy et al. (1976) comme une courte période au cours de laquelle la quantité de glaces stockées sur les continents a été au moins aussi élevée que pendant les autres stades glaciaires. Si cette corrélation était vérifiée il s'avèrerait d'une part que l'Anfatien correspond à une période de

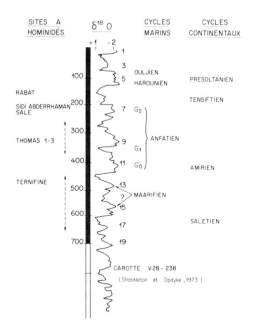

FIGURE 11.6
Tentative de corrélation entre le Pléistocène océanique et le Pléistocène marin du littoral atlantique du Maroc.

durée très importante, ce qui est en accord avec l'importance de ses dépôts sur le littoral atlantique, et d'autre part que les variations eustatiques ont joué un rôle mineur, au Maroc, par rapport aux mouvements verticaux du littoral. Il devient dès lors possible de corréler le Maarifien au stade 13 ou 15 de cette échelle. Ces corrélations permettent de proposer pour les périodes continentales du Maghreb les âges suivants (Figure 11.6):

Salétien: plus ancien que 500,000-600,000 ans
Amirien: de 480,000 à 220,000 ans (maximum transgressif Anfatien G_2)
Tensiftien: de 220,000 à 145,000 ans
Présoltanien: de 145,000 à 80,000 ans

Elles rejoignent pour l'essentiel, au moins pour les niveaux les plus récents, celles proposées par Butzer (1975) mais doivent néanmoins être considérées comme provisoires.

LE PALÉOENVIRONNEMENT DU MAGHREB AU PLÉISTOCÈNE MOYEN

Les données relatives à ce problème sont encore très incomplètes. La géomorphologie, qui a été bien étudiée au Maroc et en Tunisie, met surtout en évidence la diversité des paysages de cette époque. L'interprétation des altérations pédologiques se heurte encore trop souvent à des difficultés engendrées par une chronologie souvent douteuse. Les grands vertébrés fossiles, dont on connait l'indépendance par rapport aux divers biotopes, n'apportent que des données d'ordre très général. Les microvertébrés appartiennent pour la plupart à des espèces qui ne sont plus représentées actuellement. Enfin, les données paléobotaniques, fondamentales pour la reconstitution des paléoenvironnements, se limitent à quelques données ponctuelles, notoirement insuffisantes. Ce sont les gastéropodes du rivage atlantique qui apportent encore dans ce domaine les informations les plus complètes.

D'une façon très schématique, l'ensemble des données s'accorde pour attribuer au Pléistocène inférieur un environnement de type tropical qui n'a toutefois pas été vérifié par l'étude de la végétation de cette époque. Au Pliocène supérieur, vers 3 à 3.5 MA, des empreintes foliaires (Arambourg, Arènes, et Depape 1953) et surtout des pollens (Van Campo 1977; Sivak 1977) récoltés au lac Ichkeul suggèrent un environnement de type forêt galerie entouré d'espaces ouverts. Ce paysage est en accord avec la faune de grands mammifères du même gisement (Arambourg 1969-70) constituée d'éléments de milieu ouvert (*Camelus*, *Hippotragus*, *Oryx*, *Gazella*, et *Ceratotherium simum*) et de forêt (*Omochoerus*, *Redunca*, *Dicerorhinus*). Les micromammifères comprennent une espèce dominante méditerranéenne (*Prolagus*) associée à d'autres espèces dont la signification écologique n'est pas encore connue. Toutefois, une faune de micromammifères sensiblement contemporaine mais provenant du sud tunisien (Ain Brimba, Jaeger 1975a) a révélé une composition extrêmement distincte où dominent les espèces caractéristiques du domaine aride (Ctenodactylidé, Gerboise, Gerbille). Ce contraste entre les rivages méditerranéens et le domaine méridional représente donc une donnée très ancienne de l'histoire bioclimatique du Maghreb.

La période qui marque la transition entre le Pléistocène inférieur et le Pléistocène moyen, communément désignée sour le terme 'Villafranchien' supérieur, mais à laquelle aucun âge précis ne peut encore être attribué avec certitude (0.7 à 1.5 MA?) se caractérise par un climat chaud mais déjà très aride. Ceci se manifeste aussi bien dans l'évolution sédimentaire (Coque 1962; Coque et Jauzein 1965; Maurer 1968; Beaudet 1969) que dans la composition des faunes de rongeurs (Jaeger 1975a). C'est à cet épisode que correspond, au moins partiellement, l'optimum thermique du Messaoudien du rivage atlantique (Brébion 1973, 1976).

Le début du Pléistocène moyen se caractérise par un net refroidissement. Ainsi les Gastéropodes du Maarifien (0.48 ou 0.56 MA?) traduisent un climat tempéré. Cette dégradation de la température est confirmée par un important changement enregistré au sein des compositions des faunes de rongeurs. Dès le niveau de Ternifine apparaissent les premiers *Ellobius* et *Meriones*, originaires des steppes d'Asie centrale. D'autre part, les éléments thermophiles comme les Cténodactylidés disparaissent des provinces septentrionale et centrale (Jaeger 1969, 1975a, 1975b). La faune de grands mammifères de Ternifine évoque alors un paysage de type savane soudanaise. L'Amirien est également l'époque où les traces d'hydromorphie sont importantes (Beaudet, Maurer, et Ruellan 1967) et où le haut Rif central apparaissent les premières niches de nivation (Maurer 1968). La flore pollinique de l'Amirien inférieur est représentée par quelques taxons isolés (Van Campo, in Biberson 1961: 104): Acacia, Chenopodiacées, Amaranthacées, Caryophyllacées, Composées.

Les gastéropodes du rivage atlantique indiquent que cette période tempérée, relativement humide, a dû se prolonger jusqu'à l'Anfatien G_1. Cette dernière transgression, caractérisée par l'abondance de *Littorina littorea*, correspond au refroidissement le plus intense connu au Maghreb. C'est sans doute à cette époque que l'on peut référer le maximum d'extension des glaciers des montagnes marocaines (Raynal 1961; Maurer 1968). Paradoxalement, ni les faunes de micromammifères ni celles de grands mammifères ne paraissent subir de modification sensible dans leur composition. Cette absence de fluctuations dans la composition des faunes au cours du Pléistocène moyen représente une des caractéristiques les plus originales du Maghreb.

Toujours d'après les gastéropodes, l'Anfatien G_2 marque le retour à des températures chaudes, cette période correspondant à l'optimum thermique du Pléistocène moyen.

De nouvelles espèces de gastéropodes s'installent et prolifèrent (*Patella safiana*, *Thais haemastoma*). Toutefois à aucun moment le climat ne redevient tropical comme l'atteste l'absence totale d'altération de type ferralitique au Pléistocène moyen. Le Harounien (Rabatien) qui succède à l'Anfatien G_2 et qui correspond au premier niveau marin du Pléistocène supérieur marque le retour à des conditions climatiques plus tempérées, les faunes de gastéropodes de ce niveau ayant une composition comparable à celles du Maarifien (Brébion 1973, 1976).

Si les faunes de mammifères ne présentent pas de fluctuations liées aux variations climatiques, elles présentent par contre des modifications progressives dans leur composition. La première correspond à l'apparition progressive d'espèces caractéristiques des milieux ouverts comme les Gazelles (*G. dorcas*), le phacochère, et *Rabaticeras*, que l'on peut corréler à la disparition progressive des espèces de forêt, tels l'ours et l'*Elephas iolensis* à l'Amirien ainsi que *Cervus* au Tensiftien. Cette tendance ne se répercute dans la microfaune qu'à son paroxysme, lors de la disparition des genres *Ellobius*, *Arvicanthis*, *Praomys* associés à tout un cortège d'espèces qui dominaient les faunes du Pléistocène moyen. Cette crise majeure survient à la fin du Présoltanien ; elle correspond à une rupture brutale de l'équilibre des biocénoses et correspond à une poussée considérable de l'aridité. Cette évolution progressive vers l'aridité, malgré l'existence de fluctuations climatiques qui ne se répercutent pas dans la composition des faunes, correspond étroitement à ce que Tchernov (1968 et 1975) a décrit en Palestine. Au Maghreb cette situation découle de la géographie, les faunes locales n'ayant aucun refuge disponible au moment des paroxysmes climatiques. De ce fait, leur isolement génétique conduit à des évolutions sensiblement comparables à celles qui se produisent en milieu insulaire (Jaeger 1975a).

La deuxième tendance manifestée par les faunes de mammifères concerne la différenciation géographique en deux provinces, littorale atlantique et peut-être littorale méditerranéenne, et zone interne. La limite entre ces deux zones, dont l'existence est attestée depuis le Pliocène, se déplace progressivement vers le nord. Déjà reconnue chez les grands mammifères amiriens par Arambourg (1960) avec l'abondance des Ours et la présence d'*Elephas iolensis* sur le littoral atlantique, s'opposant à la faune soudanaise à *Elephas atlanticus* des plateaux algériens, elle est plus marquée encore chez les microvertébrés. C'est ainsi qu'on peut distinguer, au Tensiftien, une faune littorale avec *Mus musculus* et *Ellobius barbarus* s'opposant à une faune plus continentale avec *Mus* (*Leggada*) *jotterandi*, *Ellobius zimae*, *Jaculus orientalis*, et *Atlantoxerus getulus*. Cette dernière correspond à une province à végétation distincte correspondant à un milieu ouvert plus xérophytique que la présumée forêt littorale. Les paysages végétaux correspondant à ces deux domaines restent encore inconnus. Une flore foliaire des environs d'Alger, d'âge tensiftien-présoltanien, n'a livré que des éléments méditerranéens (Arambourg, Arènes, et Depape 1953). Quant à la forêt du littoral atlantique suggérée par les faunes continentales, son existence n'est étayée par aucun document paléobotanique.

En conclusion, il apparaît que le Maghreb a connu au cours du Pléistocène moyen, un climat tempéré à chaud, jamais tropical, ponctué par quelques fluctuations climatiques. La première partie a été tempérée et humide ; à un maximum de froid à l'Anfatien G_1 succède une période chaude, puis à nouveau un retour à un climat plus tempéré. Le phénomène le plus important étant le développement progressif de l'aridité, suivant une évolution strictement comparable à ce qui a été observé en Palestine.

Une dernière remarque, d'ordre biogéographique, mérite d'être signalée. L'existence d'échanges fauniques entre l'Espagne et le Maghreb a été suggérée par de nombreux auteurs. Nous avons montré (Jaeger 1975b) qu'il n'existait aucune preuve paléontologique concrète à l'appui de cette hypothèse et d'autre part que tous les immigrants eurasiatiques avaient eu une dispersion Est-Ouest, la plupart ayant en outre été découverts dans des niveaux antérieurs au Proche-Orient. Ces échanges directs Afrique-Espagne avaient été également préconisés pour expliquer la répartition des industries acheuléennes en Europe occidentale. Ils ont été récemment remis à l'ordre du jour par les travaux de Freeman (1975) et d'Alimen (1975). Certes on ne peut nier a priori la possibilité de tels échanges exclusivement limités aux seuls humains. Mais on ne peut également éliminer une autre alternative, à savoir que les cultures acheuléennes introduites sur les deux rives de la Méditerranée au début du Pléistocène moyen ont pu être refoulées vers l'Ouest, dans ces deux régions, par d'autres cultures. Seules des datations précises permettant de vérifier le synchronisme strict des stades culturels de part et d'autre de la Méditerranée représenterait, à notre avis, un argument décisif en faveur d'échanges directs de populations humaines.

LES HOMMES FOSSILES

La nature et la provenance des restes humains du Pléistocène moyen du Maghreb sont résumées dans le Tableau 11.2.

Les restes les plus anciens sont ceux de Ternifine pour lesquels nous proposons un âge compris entre 0.5 et 0.7 MA. Leur étude détaillée a conduit Arambourg (1963) à les attribuer à une variété nouvelle d'*Homo erectus*, *H. e. mauritanicus*. En effet, la structure des mandibules, les caractères généraux de la denture, la morphologie du pariétal et la trace des vaisseaux méningés justifient largement cette désignation taxonomique. L'étroite ressemblance avec les structures correspondantes de l'H 9 d'Olduvai confirme également ce point de vue. Les dimensions générales et l'estimation de la capacité cranienne (Arambourg 1963; Kochetkova 1970) permettent d'attribuer ces restes à des individus d'assez grandes dimensions.

La mandibule de la carrière Thomas I, plus récente que les restes de Ternifine et que nous rapportons à l'Amirien supérieur, a été décrite en détail par Sausse (1975). La conclusion essentielle de son étude réside dans l'attribution de ce fossile à *H. erectus mauritanicus* à cause de l'extrême similitude de la denture bien que la mandibule, quoique d'organisation identique à celles de Ternifine,

178 *Homo erectus*

TABLEAU 11.2
Inventaire des restes humains fossiles du Pléistocène moyen du Maghreb

Gisement	Nature des restes réceuillis	Nom de l'inventeur et date	1ère Publication; Etude détaillée
Ternifine	Pariétal gauche 3 mandibules (I, II, et III) 1 C. supérieure, 2 DP^3 droites, 1 DP^4 gauche, 1 M^1 droite, 1 M^2 gauche, fragment postérieur d'une M^2	Arambourg et Hoffstetter 1954, 1955, 1956	Arambourg et Hoffstetter 1954; Arambourg 1963
Thomas I	Fragment de mandibule gauche avec P$_4$-M$_3$	Beriro 1969	Ennouchi 1970; Sausse 1975
Thomas III	Fragment cranio-façial gauche avec frontal et arcade sourcilière incomplets et 4 petits fragments de la face. I^2-M^2 gauche, C-P^3-P^4-M^2-M^3 droites	Beriro 1972	Ennouchi 1972; Ennouchi 1976
Sidi Abderrahman	Fragment de mandibule droite avec M$_1$-M$_2$-M$_3$; fragment de mandibule gauche avec P$_3$	Biberson 1955	Arambourg et Biberson 1955, 1956
Salé	Calvarium avec fragment de M^3 droite, Maxillaire gauche avec I^2-M^2, moulage endocranien nat.	Jaeger et Dakka 1971	Jaeger 1975c
Rabat	3 fragments du pariétal droit, 2 fragments du pariétal gauche, 3 fragments d'occipital, maxillaire avec I^1-M^2 gauche et palais, mandibule avec I$^1_1$-M$_1$ gauche et 1$_1$,P$_3$-P$_4$,M$_1$-M$_2$ cassées et M$_3$ droites	Alenda 1933; pro parte et Marçais 1933	Marçais 1934; Saban 1975

soit plus gracile. Les restes de Thomas III, sans doute contemporains de ceux de Thomas I, sont également rapportés à *H. e. mauritanicus* (Ennouchi 1972, 1976). Les dimensions des restes dentaires[1] ainsi que leur morphologie, coincident avec les éléments correspondants de Ternifine et de Rabat. Le fragment antérieur du crâne montre un fort bourrelet supraorbitaire déprimé au niveau de la glabelle,

---

1 La M^3/ supérieure gauche figurée par Ennouchi (1976) correspond à une M^3/ droite mal orientée sur la figure 10-11. La M supérieure droite décrite comme M^1/ correspond d'avantage à une M^2/.

un frontal très bas, similaire selon Ennouchi à celui d'*H. e. erectus*. La mandibule de Sidi Abderrhaman, un peu plus récente que les restes de Thomas I et III (Arambourg et Biberson 1955, 1956), possède toutes les caractéristiques décrites chez les hominidés de Ternifine. Les restes de Rabat, décrits initialement par Vallois (1945, 1959), ont été réétudiés en détail par Saban (1975) qui a dégagé les caractéristiques essentielles de ces restes. Si les dents sont comparables à celles des autres hominidés décrits du Pléistocène moyen d'Afrique nord-occidentale, l'occipital présente, selon Saban, un mélange de caractères d'*H. erectus* et d'homme moderne, ce qui conduit cet auteur à attribuer ces restes à un présapiens, sans définir toutefois sa situation taxonomique précise. Les caractères modernes, essentiellement localisés à la partie extérieure de l'occipital, sont: l'absence de torus occipital, l'absence de chignon, et le contour arrondi de la face externe de l'occipital. La face interne du même os présenterait, selon cet auteur, des caractères plus archaiques, telle la position de l'endinion, situé un centimètre sous l'inion et la forte disymétrie des structures. Les restes de Salé, décrits sommairement par l'auteur (Jaeger 1975a), constituent les documents les plus complets relatifs aux hominidés du Pléistocène moyen de Maghreb. Ils proviennent d'un horizon dunaire daté d'environ 220,000 ans, sensiblement contemporain du gisement qui a livré la mandibule de Sidi Abderrahman. Il serait donc plus ancien que l'homme de Rabat que plusieurs auteurs situent, par des raisonnements distincts, entre 145,000 et 200,000 ans (Biberson 1961, Saban 1975). Ceci est confirmé par la tentative de corrélation et de datation proposée dans le présent travail.

Le crâne de Salé qui appartenait à un adulte, possède un grand nombre de caractères d'*H. erectus* dont il convient d'énumérer les plus importants (Figure 11.7):

1 Capacité crânienne faible, comprise entre 930 et 960 cm^3;
2 dimensions générales réduites, comprises pour la plupart dans la limite de variation des *H. erectus* classiques;
3 platycéphalie (Figure 11.8);
4 frontal caréné avec forte constriction postorbitaire et bregma saillant. Sa morphologie suggère l'existence d'un torus supraorbitaire puissant;
5 plus grande largeur située à la base du crâne;
6 torus occipital;
7 bord supérieur rectiligne du temporal;
8 base du crâne très aplatie;
9 apophyse mastoide de section triangulaire avec faces postérieures aplaties dans le plan nuchal fortement inclinées vers l'intérieur;
10 rocher et tympanique déterminant un angle de 140°;

180 *Homo erectus*

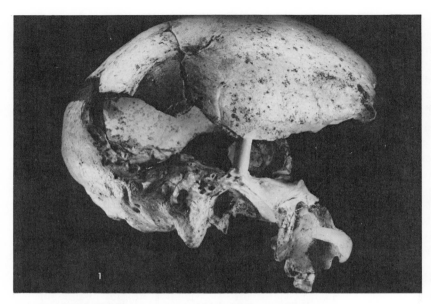

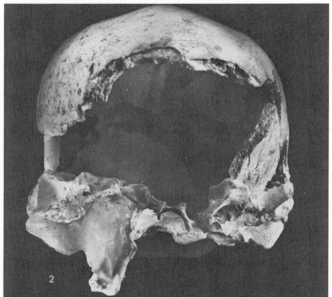

FIGURE 11.7
1: Crâne de Salé (Maroc) en vue latérale droite (6.1). 2: Crâne de Salé en vue antérieure (.6:1).

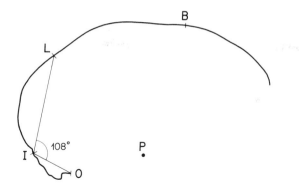

FIGURE 11.8
Diagramme saggital médian du crâne de Salé (.4:1).

11 fort épaississement de l'os dans la région astérique;
12 denture robuste, caractéristique par ses dimensions et sa morphologie, d'*H. erectus* et identique à celles de l'*H.* de Rabat et de Thomas III;
13 empreintes des vaisseaux méningés moyens, comparables à celles d'*H. e. mauritanicus* (Figure 11.9). La branche temporale inférieure est modérément développée, mais les branches bregmatiques et obéliques sont simples, peu ramifiées.

D'autres caractères, moins nombreux, s'écartent de la liste classique des *H. erectus* et évoquent un degré d'organisation plus moderne. Les pariétaux, aux dimensions et aux proportions identiques à ceux des *H. erectus* classiques, présentent une courbure transverse importante marquée par l'ébauche de bosses pariétales dont le développement confère au crâne un contour presque rectangulaire en vue postérieure. Le basi-occipital et le basi-sphénoide ont des proportions modernes. Les apophyses vaginales sont bien développées, mais il n'y a pas de trace de styloides.

L'occipital possède un cachet très particulier. Sa hauteur totale est faible, l'écaille supérieure est peu élevée, séparée de l'écaille inférieure par un torus occipital faiblement marqué, aplati et développé en hauteur qui s'estompe latéralement. L'inion est disposé 2 cm en-dessous de l'opisthocranien. Les lignes occipitales supérieures sont irrégulières. Les lignes occipitales inférieures correspondent à un puissant bourrelet de contour rectangulaire qui délimite vers le haut de profondes dépressions correspondant vraisemblablement aux insertions des muscles petits droits postérieurs. L'angle opisthion-inion-lambda est de 108 degrés (Figure 11.8). La face interne de l'occipital (Figure 11.10), parfaitement

182 *Homo erectus*

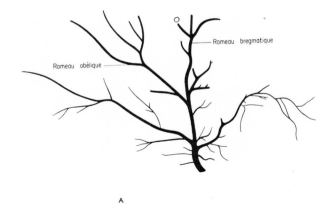

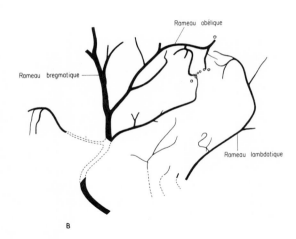

FIGURE 11.9
Traces des vaisseaux méningés relevés sur le moulage endocranien du crâne de Salé. A: côté droit; B: côté gauche.

conservée, montre les empreintes des lobes occipitaux et des hémisphères cérébelleux séparés par les sinus veineux gauche et droit qui déterminent entre eux un angle de 90 degrés. Les empreintes du cerveau frappent par leurs petites dimensions, bien en accord avec la faible capacité crânienne. Le sinus veineux droit est plus développé que le gauche. Ces sinus sont disposés au niveau des

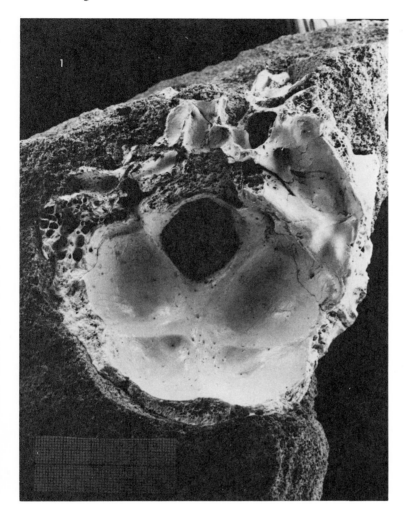

FIGURE 11.10
Vue endocranienne de la base du crâne de Salé, encore inclus dans sa gangue d'éolianite (.75:1).

lignes occipitales supérieures. L'endinion se situe environ un centimètre en dessous du niveau de l'inion. Par beaucoup de ses caractères, l'occipital de Salé se rapproche de celui de Rabat: son contour externe arrondi, son torus réduit sinon totalement absent comme à Rabat, les positions respectives de l'inion et de l'endionion. Il en diffère par l'agencement des sinus veineux, mais ce caractère est très variable chez les hommes actuels. Ces observations nous conduisent à penser que les différences énumérées peuvent être attribuées d'une part à l'âge juvénile de l'homme de Rabat et d'autre part à la variabilité individuelle. Ce point de vue qui conduit à intégrer l'homme de Rabat au même taxon que celui de Salé est également appuyé par la forte ressemblance des dentures supérieures de ces deux hominidés.

Le problème le plus immédiat posé par ces restes humains du Maghreb, largement répartis dans l'intervalle du Pléistocène moyen, est celui de leurs relations phylétiques. Des liens phylétiques directs peuvent relier d'une part la population de Ternifine au groupe Sidi Abderrhaman-Salé-Rabat et d'autre part ce dernier groupe aux hommes modernes fossiles du Maghreb (Irhoud 1 et 2, Tanger, Haua Fteah, Temara, Dar es Soltane). Il convient de rappeler en effet que de nombreux auteurs, Arambourg en premier (1955, 1963) considèraient que tous les hominidés du Pléistocène moyen de cette région constituaient une lignée évolutive, les plus récents dérivant de ceux de Ternifine. Cette interprétation fut défendue également par Howell (1960), alors que Cabot Briggs (1968), Tobias (1968a), et Saban (1972) prolongeaient même cette lignée évolutive jusqu'aux hommes modernes du Maghreb. Vallois (1945) a décrit l'homme de Rabat comme un prénéanderthalien distinct de ceux contemporains d'Europe occidentale alors que Saban (1975) voit en l'homme de Rabat un présapiens à caractères hérités des ancêtres 'archanthropiens' mais à évolution parallèle à celle des 'présapiens' d'Europe et contemporain d'un 'pithécanthrope attardé,' l'homme de Salé. Nous pensons avoir apporté suffisamment d'arguments pour démontrer d'une part que les restes de Sidi Abderrhaman-Rabat-Salé appartiennent à un même type humain rattachable au groupe des *Homo erectus*, malgré quelques caractères modernes inattendus. Ce sont précisément ces derniers caractères dont l'interprétation reste particulièrement délicate. Compte tenu de la faible variabilité des *Homo erectus* d'Asie et d'Afrique les plus anciens (*H. erectus erectus, H. e. pekinensis*; H 9 Olduvai, KNM-ER 3733) on peut raisonnablement admettre que les hominidés de Ternifine avaient une région occipitale similaire, avec un puissant torus occipital et un angle occipital très aigu. On connaît par ailleurs la faible courbure transverse du pariétal de Ternifine ce qui nous conduit à admettre que s'il y a eu évolution au Maghreb, celle-ci s'est manifestée uniquement par le développement des bosses pariétales et la rotation de l'occipital, ce qui en 0.5 MA environ, représente un taux de transformation morphologique dérisoire. Cependant, le fait que la population de Ternifine soit de grande taille, avec une large capacité crânienne, vient s'opposer aux dimensions réduites des individus de Salé et de Thomas III, à faible capacité crânienne. De ce fait, il nous parait pour le moins prématuré, dans l'état actuel de nos connaissances, de faire dériver le groupe de Salé directement de celui de Ternifine.

Le problème des relations avec les hommes modernes fossiles du Maghreb est encore plus délicat. Les hommes d'Irhoud I et II, à caractères modernes et à forte capacité crânienne (Ennouchi 1962a, 1962b, 1966, 1968, 1969a, 1969b) n'ont été décrits que sommairement. Les différences sont considérables avec Salé, sauf en ce qui concerne la robustesse de la denture et la position haute de la largeur bipariétale maximale. D'autre part, l'âge très récent du groupe humain de Salé et la persistance de si nombreux caractères d'*H. erectus* semble devoir faire appel à une vitesse d'évolution et à des transformations considérables pendant un intervalle de temps très court, transformations dont l'existence ne peut encore être établie avec certitude. Comme nous l'avons indiqué clairement dans un chapitre antérieur, nous ne pensons pas qu'une quelconque continuité dans le développement de l'industrie lithique puisse représenter un argument valable dans l'étude de l'évolution humaine pour attester l'existence d'une continuité génétique. La mandibule de Temara (Vallois et Roche 1958; Biberson 1961) avait été initialement décrite comme associant des caractères modernes à des caractères primitifs, ces derniers évoquant ceux des autres hominidés du Maroc atlantique, plus anciens. Or, la réexploitation du site fossilifère a permis à Roche et Texier (1976) de démontrer que cette mandibule provenait en réalité de niveaux ayant livré une industrie lithique Atérien supérieur, donc considérablement plus récents que ce qui avait été admis antérieurement. Ces nouvelles fouilles ont livré d'autres restes humains, dont une partie postérieure de crâne attribuée par Ferembach (1976) à *Homo sapiens sapiens*.

S'il n'apparait pas de continuité entre le groupe d'*H. erectus* de Salé et les hommes modernes fossiles du Maghreb, il convient de réexaminer les caractères modernes de ces *H. erectus* et de tenter d'en dégager la signification. Parmi ces caractères modernes énumérés par l'auteur en 1975, figurent l'absence de chignon occipital, la réduction du torus occipital, l'ouverture de l'angle occipital, et le développement des bosses pariétales. Le premier de ces caractères ne suffit pas à éliminer le crâne de Salé des *H. erectus* puisqu'il présente le même contour postérieur en vue supérieure que le crâne d'*H. e. erectus* VII décrit par Sartono (1968). La configuration de l'occipital de Salé ne peut être considérée comme fondamentalement distincte de celle d'un *H. erectus* classique.

Nous pensons que le faible développement du torus associé à une légère ouverture de l'angle occipital suffit à provoquer un décalage vers le bas des lignes occipitales supérieures et de l'inion. Ainsi les caractères modernes du crâne de Salé se limiteraient à deux caractères principaux, le développement des bosses pariétales et une ouverture de l'angle occipital. Le développement de ces deux caractères ne peut être interprété qu'avec circonspection. Il peut s'agir d'un gradualisme phylétique (ou chronocline) à partir d'une population ancestrale (Ternifine?) plus primitive. Si cette hypothèse se révélait exacte elle confirmerait pleinement les conceptions d'Eldredge et Gould (1972), exposées récemment de façon plus complète (Gould et Eldredge 1977), selon lesquelles le gradualisme phylétique n'aurait qu'un rôle mineur dans l'évolution, qui se ferait essentiellement par spéciations rapides aléatoires, suivi d'un tri drastique exercé par la sélection naturelle sur les espèces nouvelles ainsi apparues.

L'apparente continuité des structures au sein d'une lignée évolutive ne serait ainsi qu'un artefact, l'analyse détaillée mettant en évidence de très nombreuses discontinuités. Comme le montrent ces auteurs, il convient de connaître non seulement des documents d'âges distincts mais aussi de connaître la part qui, à chaque niveau, revient à la variation géographique. On est donc loin de pouvoir démontrer l'existence de lignées anagénétiques vraies dans la ligne humaine.

Le problème du déterminisme de l'apparition des caractères modernes du fossile de Salé ne peut être abordé qu'avec la plus grande circonspection. S'il est vraisemblable que le dimorphisme sexuel (il s'agit peut-être d'un individu femelle) a pu exagérer le développement de ces caractères modernes, sans aucun doute génétiques, il est toutefois difficile de les attribuer à une quelconque sélection puisqu'ils ne sont pas accompagnés d'une apparente augmentation de la capacité crânienne. Reste alors la dérive génique, plus vraisemblable, ou la néotenie, peu justifiable.

S'il ne résoud pas tous les problèmes afférents au devenir des *H. erectus*, le crâne de Salé complète de façon considérable les données relatives à l'évolution humaine au Maghreb. Trois points sont particulièrement mis en évidence:

1 Les caractères primitifs de la denture et de l'appareil masticateur persistent pendant une durée considérable chez les Hominidés fossiles du Maghreb sans présenter aucun signe d'évolution sensible. Rappelons en effet que la mandibule de l'homme de Temara, longtemps attribuée à un pithécanthropien, et qui a été récemment assignée à un *Homo sapiens sapiens* (Ferembach 1976), possède encore plusieurs caractères archaïques, son appartenance à un homme moderne n'étant trahie que par une ébauche d'éminence mentonnière. Cette observation confirme l'opinion de Pilbeam (1975) selon laquelle la denture et l'appareil masticatoire ne suffisent pas pour définir les relations phylétiques chez l'homme. L'explication d'une telle stabilité devrait être recherchée dans le mode de vie et d'alimentation de ces hominidés.

2 La détermination taxonomique des hommes du Pléistocène moyen doit être faite avec prudence, compte tenu de la relative indépendance des différentes structures les unes avec les autres. Ainsi la seule découverte de l'occipital de Salé aurait peut-être conduit à son attribution à *Homo sapiens* ou à un présapiens,

comme cela a été suggéré par Saban (1975) à la suite de son étude des restes humains de Rabat. Ceci pose tout particulièrement le problème de l'attribution taxonomique des restes humains de Swanscombe et de Vértesszöllös en Europe.
3 La variabilité chez les hommes fossiles du Pléistocène moyen est au moins aussi importante que chez l'homme moderne. Entre l'homme de Rabat et de Salé plusieurs différences attestent l'importance de cette variabilité. C'est le cas pour le développement du torus occipital, faible à Salé, absent à Rabat et de l'agencement des sinus veineux gauche et droit, très disymétriques chez l'homme de Rabat, perpendiculaires chez l'homme de Salé.

CONCLUSION

En conclusion, il apparaît que jusqu'à environ 200,000 ans vivait sur le littoral du Maroc une population d'*Homo erectus* possédant quelques caractères modernes qui la distinguent des autres populations connues. Ces hominidés de la fin du Pléistocène moyen du Maghreb dérivent directement ou indirectement de ceux, plus anciens, de Ternifine, dont la morphologie devait se rapprocher d'avantage de celle des *H. erectus* les plus anciens. Compte tenu du grand nombre de caractères primitifs rencontrés chez des hommes aussi récents, il est peu vraisemblable qu'ils représentent les ancêtres des *Homo sapiens* qui leur ont succédé dans cette partie du continent africain. Toutefois leur évolution illustre une incontestable tentative vers la sapientisation démontrant que les *Homo erectus* possédaient dans leur patrimoine génétique la possibilité de se transformer en homme moderne. L'ensemble des données, y compris les plus récentes (Clarke 1976), tendent à montrer que de telles tentatives n'ont pas seulement été limitées à l'Europe, l'Afrique du Nord, ou l'Asie du Sud-Est, mais qu'elles correspondent à une constante de l'histoire des *Homo erectus*.

12

G.P. RIGHTMIRE

Homo erectus at Olduvai Gorge, Tanzania

A la suite des découvertes récentes au lac Turkana, il apparaît clairement qu'*Homo erectus* a eu une longue histoire en Afrique orientale, qui a commencé dès avant le début du Pléistocène moyen. Des restes d'*Homo erectus* ont également été recueillis à Olduvai, en Tanzanie, où des fragments d'une lourde boîte crânienne ont été découverts par L.S.B. Leakey en 1960. On étudie actuellement la partie de crâne ainsi reconstituée et d'autres documents tirés du niveau IV des gorges d'Olduvai. La présente étude ne contient que des observations préliminaires sur l'importance des hominiens d'Olduvai, HO 9 et HO 12; des descriptions détaillées apparaîtront dans d'autres ouvrages.

Recent discoveries at Lake Turkana (Walker, this volume) make it clear that *Homo erectus* had a long history in East Africa, beginning well before the onset of the Middle Pleistocene. Remains of *Homo erectus* are known also from Olduvai Gorge in Tanzania, where pieces of a heavy braincase were recovered by L.S.B. Leakey in 1960. This partial cranium and other materials from Bed IV at the gorge are now under study. Detailed descriptions will appear elsewhere (Rightmire 1979), and only brief preliminary comments on the significance of the Olduvai hominids, OH 9 and OH 12, will be presented here.

Hominid OH 9 was found at site LLK in the side Gorge. Dating of these upper Bed II deposits has been difficult, as only one reliable K/Ar date has been obtained from a marker tuff in the lower part of Bed II. Age estimates are based instead on magnetic polarity, and extrapolation from stratal thicknesses between the top of the Lemuta Member and the magnetic reversal known to have occurred at 0.7 million years in middle Bed IV suggests an age of 1.1 million years for the top of Bed II (Hay 1976). Several quartz flakes and other tools were recovered from excavations at the site, but the few utilized pieces are not diagnostic, and the hominid cannot be associated directly with any of the Olduvai cultural traditions.

190 *Homo erectus*

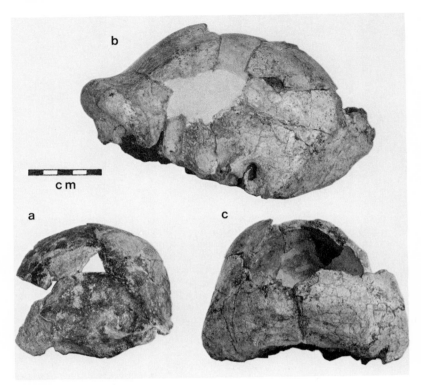

FIGURE 12.1
a: rear of the Olduvai Hominid 12 braincase; b: lateral view of the Olduvai Hominid 9 cranium; c: occipital view of the Olduvai Hominid 9 cranium.

The partial braincase includes the supraorbital structures, occiput and much of the base, although the facial skeleton is missing. Some plaster has been used to reconstruct parts of the frontal and parietals, and small quantities of matrix still adhere to the underside of the specimen. However, the remaining bone is well preserved and deformation of the cranium is minimal. A striking feature is the heavy supraorbital torus, which is especially thick above the nose. The surface bone is vermiculate, and the orbital rims sweep downward and back to accentuate the shelf-like form of the brows, which appear to overhang the upper face. The supratoral region is hollowed laterally to produce a shallow sulcus on each side, and there is no indication of frontal keeling in the midline.

In lateral aspect, the vault is low and flattened. The temporal line is raised on the parietal but fades inferiorly to reappear as a weak suprameatal crest before merging with the root of the zygomatic process. This process is heavily built, with a broad superior surface that falls steeply into the infratemporal fossa. On the occiput, the superior nuchal lines traverse a rounded torus and can then be traced onto the posterior faces of the pyramidal mastoid processes. The nuchal area of the occiput is continuous with the broad, flattened posterior portions of the mastoids, and maximum cranial breadth falls at the mastoid crests. The glenoid fossa is very deep. The tympanic plate constitutes the rear wall of the fossa and is oriented nearly vertically. A strong vaginal process is apparent, and a circular depression marks the location of the styloid process, the broken root of which is clearly present. The interior of the braincase has been partly reconstructed, and some areas are badly damaged. However, some information concerning the base can be obtained, and Holloway (1975) has measured the endocranial volume of OH 9 as 1067 cm^3.

Fragments of the smaller OH 12 cranium were picked up on the surface at site VEK in deposits of Bed IV. The Brunhes-Matuyama magnetic reversal can be located no lower than the base of Tuff IVB within this bed, so an age span for the deposits from about 0.8 million years to 0.6 million years has been suggested (Hay 1976). M.D. Leakey (1975) reports the persistence of a Developed Oldowan industry at a series of sites on the south side of the gorge, while Acheulean artifacts have been recovered at several levels in Bed IV and from at least one level in the overlying Masek Beds. Cultural associations of OH 12 are unknown, although other skeletal remains attributable to *Homo erectus* have been found with an Acheulean assemblage at site WK in upper Bed IV (M.D. Leakey 1971; Day 1971).

The back of the braincase has been reconstructed from many small pieces, some of which are badly weathered. The upper part of the occipital squama is intact, and most of the lambdoid margin is preserved on the right side. An external protuberance is present but not so prominent as in OH 9, and a true occipital torus is not developed. Parts of both parietals are attached, and in rear view the vault is rounded. There is no sagittal keeling, and the temporal lines are very faintly marked. Close to the midline, a long oval depression is present in the bone of the right parietal, and this damage must have been inflicted before the skull was mineralized.

Another composite fragment is made up of small bits of parietal bone united with part of the frontal at bregma. Unfortunately this piece cannot reliably be joined to the rest of the vault, and the frontal bone is preserved for only a short distance anteriorly. More of this bone is represented by a fragment from the upper margin of the right orbit. A short section of the supraorbital torus is intact, and this is much more lightly constructed than the brow of OH 9. Portions of both temporals are too small to provide much anatomical detail, but the mastoid process is short and nipple-like. A left maxilla is also badly broken, but the damaged roots of the first premolar through second molar are still in place.

Further comparative study will be needed in order to assess similarities of these Olduvai remains to other *Homo erectus* assemblages from East and North Africa. Work on the Olduvai and Koobi Fora specimens is underway, but direct comparison of crania with the Ternifine individuals principally represented by mandibles is problematical. Although there is substantial anatomical variation and also uncertainty regarding dates, it may be best to refer much of this African material to one taxon. However, it is unlikely that fossils sorted in this fashion will provide adequate samples of any single population actually existing in the Pleistocene. Some mixing of individuals drawn from regionally distinct groups living at different time periods is inevitable. Relationships of African *Homo erectus* to other approximately contemporary populations of the Old World and to later *Homo sapiens* are also obscure. There is currently support both for models of gradual evolution within *Homo erectus* lineages and for views advocating more rapid change accompanied by large-scale replacement and hybridization. The evolution of species is a complex process, and the sparse mid-Pleistocene hominid record hardly provides an ideal context in which to test phylogenetic hypotheses. But the remains from Olduvai and other African localities should shed some light on the range of *Homo erectus* variation and on links between this species and archaic *Homo sapiens*.

13

ALAN WALKER

The Koobi Fora Hominids and their bearing on the origins of the genus *Homo*

La région du Turkana oriental a grandement contribué à l'étude de l'évolution de l'homme, car on y a trouvé des hominidés fossiles et d'autres indices permettant de reconstituer son environnement. D'importance capitale, l'âge du tuf KBS est encore incertain. Selon qu'il date de 2.4 à 2.6 ou de 1.6 à 1.8 millions d'années, il faudra dans une certaine mesure repenser le chaînon évolutif qui aboutit à l'*Homo erectus*.

Il est ensuite question des petits échantillons du dossier fossile, qui diffèrent beaucoup entre eux. Trois variantes morphologiques du crâne se présentent dans le membre supérieur de la formation du Koobi Fora. Parmi les cinq hypothèses avancées, la plus plausible serait que ces trois formes représentent trois espèces d'hominidés distinctes.

The sample of early hominid fossils from Africa has increased dramatically over the last decade. The Koobi Fora Research Project, under the leadership of Richard Leakey and Glynn Isaac, has been a major factor in creating collections, not only of hominids, but also of other animal groups. These new collections have been supported by a wealth of detailed studies on the geology, geochronology, and taphonomy and have their value enhanced by coming from the same depositional basin as a number of very important early archaeological sites.

The publication of the new fossils has been in two forms. First there are Richard Leakey's short notes and commentaries (Leakey 1970, 1971, 1972, 1973a, 1973b, 1974, 1976a; Leakey and Walker 1976) and second there are fairly detailed descriptions by Leakey and his colleagues (Leakey et al. 1971, 1972; Day and Leakey 1973, 1974; Leakey and Walker 1973; Leakey and Wood 1974a, 1974b; Day et al; 1975, 1976; Leakey, Rose, and Walker 1978). The reasons for this manner of presentation are as follows. When it became clear that the Koobi Fora sediments were very rich in hominid fossils it was decided that as well as presenting the new material as quickly as possible in the form of

194 *Homo erectus*

announcements, we would be providing a service to other paleoanthropologists by delaying formal, monographic treatment in favor of detailed descriptions, photographs, and measurements. In this way we felt we could provide a source of data for those who did not have ready access to original specimens. Thus others would be able to contribute to discussions about the fossils, since they would not have to wait the many years it takes to produce a monographic, comparative account. Those of us involved in making these basic descriptive accounts did so knowing that the more discussion generated among other paleoanthropologists, the better would be our own final accounts when they came to be written. However, the initial warm response to this program has cooled somewhat; many of our colleagues now say that they would prefer more interpretative accounts. This is, in fact, more difficult than it would seem. Not only is it time consuming to compile a fully documented comparative account with interpretations, but there are many other new hominid fossils that have been found at other sites that would have to be considered.

In this paper I am going to deal first with the Koobi Fora hominids and then attempt an interpretation that takes into account fossils from elsewhere. A history of the Koobi Fora Research Project and an introduction to the hominid fossils and their context can be found in Leakey and Leakey (1978).

GEOLOGICAL SETTING

In order to use the hominid fossils in any evolutionary scheme (other than a purely cladistic one to establish relationships), their placement in time must be established. Many geological and geochronological studies have been carried out in the area to the east of Lake Turkana that enable us to place the fossils in their correct geological context (Behrensmeyer 1970; Vondra et al. 1971; Bowen and Vondra 1973; Johnson 1974; Vondra and Bowen 1976, 1978; Findlater 1976, 1978a, 1978b; Cerling 1976; Johnson and Raynolds 1976; Cerling et al. 1975). In addition, Behrensmeyer, Bowen, and Findlater have recorded microstratigraphic sections at nearly all of the hominid localities in order to determine the environment of sedimentary burial (Behrensmeyer 1975, 1976; Findlater 1978a, 1978b; Leakey, Leakey, and Behrensmeyer 1978). This task has been undertaken in an attempt to make generalizations about paleoenvironments from burial environments.

Potassium-argon age determinations have been made on some of the tuffs in the basin (Fitch and Miller 1970, 1976; Fitch et al. 1974, 1976, 1978; Curtis et al. 1975), and the results from one particular tuff have caused a celebrated minor controversy. Fission track dating of zircons from the same tuffs have been carried out by Hurford (1974) and Hurford, Gleadow, and Naeser (1976). Paleomagnetic stratigraphic studies have also been used in an attempt to extend the isotopic age results into the main sediment body (Brock and Isaac 1974, 1976). Although the broad outlines of the geology are relatively simple, the wealth of information available on detailed problems is overwhelming. The account given here is necessarily simplified and incomplete.

About 800 km^2 of sediments are exposed in a wide coastal strip from the Ethiopian border in the north to south of Allia Bay (see Figure 13.1). The sediments are about 300 m in thickness and are draped over older volcanic rocks such that, on the whole, they dip away from both the Suregei escarpment in the east and the Kokoi volcanic horst that lies between the Ileret and Koobi Fora regions. The sediments vary from fluviatile to lacustrine, with channel and floodplain deposits predominant inland and lacustrine deposits predominant towards the lake.

The whole sediment body is a complex interdigitating series of alternating fluviatile and lacustrine sediments brought about by alternating lake regressions and transgressions. There were periods when the land was being built out into the lake by delta and coastal plain incrementation and there were times when the lake transgressed and lacustrine deposition took place far to the east. These oscillations are long term and relatively gradual and are probably representative of times when the lake was externally drained and thus had a stable water level. Superimposed on the oscillatory cycle are more sudden events where lake levels changed rapidly and erosion surfaces were either cut into the previously laid down sediments or suddenly inundated. These events probably took place when the lake level was low and the lake was internally drained, as it is today. Thus at any particular time the sedimentary environments from east to west were, in order, alluvial valley plain, alluvial valley coastal plain and/or delta plain, lacustrine high energy, and lacustrine low energy. However, at any particular time the boundaries of these environments would not necessarily be at the same place as at another. For instance, at the maximum of the transgression following the post-KBS tuff erosion surface, the shoreline in the Koobi Fora region was well inland towards the escarpment, whereas at Okote tuff times it was nearly where it is today. Put another way, if a core were taken from a drilling that went vertically through the sediments, then alternate thicknesses of fluviatile and lacustrine sediments would be found, together with their intermediates. These would be cut across in places by erosion surfaces that were in turn filled by sediments deposited during the next lake transgression.

At infrequent intervals, the lakes that fed the area would become choked with volcanic ash from some as yet unknown source or sources. These ashes have been deposited in both river channels and/or floodplains. If they were not destroyed by subsequent erosion, they remain today as clear tuffaceous marker

196 *Homo erectus*

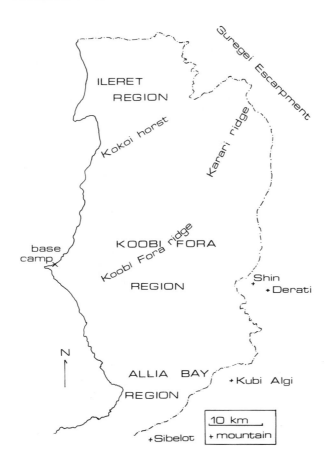

FIGURE 13.1
Map of the Koobi Fora region showing the major geographical features. Lake Turkana is to the west of the coastline.

horizons for geological correlation and as the source of material for radiometric age determinations. They are isochronous horizons and as such are extremely important, for in the complexities of the rapid lateral facies changes that occur throughout the rest of the sediment body, such horizons are not easy to distinguish. Detailed stratigraphic sections which place the hominid fossils in their exact positions in a stratigraphic section for a particular area have been published (Leakey, Leakey, and Behrensmeyer 1978). It cannot be overemphasized that these sections will give only relative positions of the hominids within that particular section. An estimate of the temporal positions cannot be made from sediment thickness or position between dated tuffs.

The sediments are exposed in three clearly defined major regions: Ileret, to the north of the Kokoi horst; Koobi Fora, to the east of the base camp as far as the escarpment; and Allia Bay, to the south. Each region is divided into smaller areas for convenience of reference. The areas are bounded by natural features such as sand rivers, patches of vegetation, etc. The areas are shown in Figure 13.2. The major tuffs used for correlation both within and between areas are given in Figure 13.3. The most important correlations are those between the Ileret and Koobi Fora regions. The correlation between the Chari and Karari tuffs is considered to be satisfactory. The Ileret Lower and Middle tuffs, as a complex, are correlated with the Okote complex (that includes the Koobi Fora tuff). The KBS tuff, formally named in area 105, is satisfactorily correlated through the Karari region to Ileret, but is less safely correlated, it appears, towards Koobi Fora. White and Harris (1977) have questioned this together with the correlation of the underlying tuffs along the Koobi Fora ridge areas. Their doubts have arisen from their studies of suid evolution in Africa. However, the exact placement of the KBS tuff in these areas probably has bearing on only two hominid fossils, these being very fragmentary specimens.

Radioisotopic age determinations on feldspar crystals from the tuffs show that the Chari/Karari tuffs are about 1.2 million to 1.3 million years old. The overlying Guomde Formation is thought to be, on paleomagnetic grounds, about 700,000 years old. The Middle/Lower-Okote tuff complex is between 1.5 million and 1.6 million years old. The KBS tuff complex is either about 2.4 million years (Fitch et al. 1976) or between 1.6 million and 1.8 million years (Curtis et al. 1975) [see pp. 229-30 of this volume]. The Tulu Bor tuff is a little over 3 million years old.

The sediments between the KBS and Chari/Karari tuff complexes are known as the Upper Member of the Koobi Fora Formation (Bowen and Vondra 1973). Those below the KBS and above the Suregei tuff complex are the Lower Member of the Koobi Fora Formation. The Kubi Algi Formation lies below the Suregei tuff complex. For the purposes of dealing with the early hominid fossil record, the units to be considered are: the Guomde Formation (only at Ileret); the upper and lower parts of the Upper Member of the Koobi Fora Formation; and the upper part of the Lower Member of the Koobi Fora Formation. Only tooth fragments have been found in the lower part of the Lower Member of the Koobi Fora Formation. Basic stratigraphic and dating information is compiled in Figure 13.3.

198 *Homo erectus*

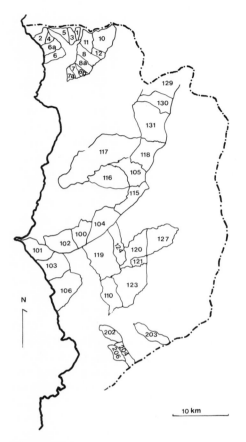

FIGURE 13.2
Collection areas in the region.

What evidence can be brought to bear on the problem of the date of the KBS tuff? It is clear that both the Cambridge and Berkeley laboratories have been trying their best to resolve the issue. The two groups are using slightly different techniques and many of the arguments deal with complex geophysics and geochemistry. Whatever the real age of the KBS tuff may turn out to be, it must be remembered that the task of the geochronologist is to date the age of the rock (in this case the age of last heating of the feldspars). Then the age of the rock still has to be related to the sedimentary and biological events for which an age determination is needed. It is clear, then, that paleontologists have nothing to

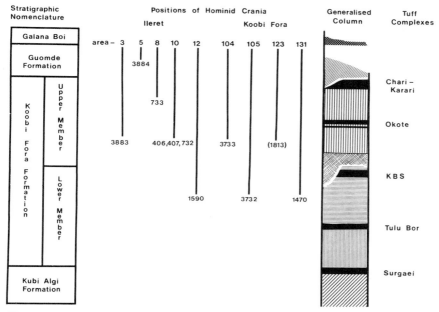

FIGURE 13.3
Stratigraphic nomenclature and relative positions of the hominid fossils at East Turkana.

offer the geophysicists to help resolve what are basically geophysical and geochemical problems. Paleomagnetic stratigraphy studies (Brock and Isaac 1974) that seemed to corroborate the earlier dates are not as straightforward as had been thought (Brock and Isaac 1976), nor, apparently, is the polarity time-scale itself (Brown and Shuey 1976). Fission-track determinations give an age of about 2.4 million years (Hurford, Gleadow, and Naeser 1976) and thus support the older date. White and Harris (1977), in their study of pig evolution, suggest that the fossil pigs would support the younger age, that is, 1.6 million years to 1.8 million years. Dr P.G. Williamson of Bristol University is using the molluscs. His results may be of great help, since the sample sizes of invertebrate fossils can be large and the results, therefore, more reliable. As far as paleoanthropologists are concerned, then, we must wait for the solution.

THE HOMINID SAMPLE

Most of the hominids are surface finds. That is, they were found washing out of the sediments. The question of provenience must be raised in such cases, because

200 Homo erectus

TABLE 13.1
Representation of parts of the skeleton in the East Turkana hominids; postcranial parts are either complete or incomplete

Cranial and axial		Upper limb		Lower limb	
Crania with teeth	7	Scapulae	1	Pelvis	2
Crania without teeth	2	Humeri	9	Femora	23
Calvaria only	1	Radii	5	Tibiae	13
Mandibles with teeth	19	Ulnae	4	Fibulae	4
Mandibles without teeth	26	Metacarpals	2	Tarsals	4
Maxillae with teeth	2	Phalanges	2	Metatarsals	4
Maxillae without teeth	3			Phalanges	6
Cranial fragments	10				
Isolated teeth or groups of teeth	31				
Vertebrae	3				

direct evidence of stratigraphic level is lacking. In very nearly all the cases, the fossils can be shown to have come from a nearby horizon. The sediments are nearly horizontal and the elevations of the erosion gulleys in which the fossils are found are not great. The chances of a mistake being made in the stratigraphic level, then, are slight, and the limits of uncertainties are recorded. The Koobi Fora sample has suffered from many taphonomic biases (see Behrensmeyer 1975), among them being differential preservation of body parts. Table 13.1 is a compilation of the Koobi Fora sample by body part. By far the greatest representation is teeth, followed by mandibles or mandibular fragments. This is the usual case in vertebrate fossil collections because the teeth and jaws are better able to resist carnivore and scavenger activity, as well as the destructive processes of sedimentary burial and exhumation. Postcranial elements are not common and there is a total lack of many parts. Associated cranial and postcranial material is rare and associated teeth, jaws, and postcranial elements are rarer.

To give some idea of the size of the Koobi Fora sample, we can ask the following questions. What fraction of the original hominid population is now represented in the collection? We can take a two-million-year period (about the span covered by the sediments) and use a generation length of twenty years. For hominids that had a population density as low as that of hunting dogs (*Lycaon pictus*) there would have been only eight per generation over the area (0.01 per km^2 [Kruuk 1972]). For hominids that had a population density as high as that of baboons (*Papio anubis*) there would have been 8000 per generation over the area (10 per km^2 [DeVore and Hall 1965]). Calculating for the 100,000 generations over the two-million-year period, the 150 individuals in the Koobi Fora collection represent only two ten-thousandths to two ten-millionths of the original population. We can further ask, what is the fraction represented by skulls, or at least good cranial specimens? The answer, following the same calculation, is between one hundred-thousandth and one hundred-millionth. This calculating has been carried out as though there were only one species of hominid. If we were to split the collections into several species, then the fractions would be much smaller. What this shows, regardless of how accurate the figures might be, is that the sample of past populations that we work with is miniscule. We might as well take as representative, in the worst instance, two individuals from the population of the United States.

I am not dealing here with fragmentary specimens or postcranial specimens, but only relatively complete crania. One of the examples, KNM-ER 1813, will demonstrate my reluctance at this stage to enter into taxonomic conjecture over fragments of mandibles, isolated teeth, or even whole mandibles with teeth. In dealing with the Koobi Fora cranial specimens I am going to claim that I can see three cranial variants (Leakey, Leakey, and Behrensmeyer 1978; Walker and Leakey 1978) and that there is discontinuous variation throughout the sample. Before some of my readers close the book now, I will later agree that there is a possibility that I am quite wrong about this.

The following are the good cranial specimens from Koobi Fora.

KNM-ER 406 (Leakey 1970; Leakey et al. 1971)
This is a fine specimen of a large adult cranium lacking the tooth crowns and incisor roots, some of the bone of the facial skeleton, and the tips of the mastoid processes. This is clearly an individual very close in morphology and size to the hyper-robust, crested *Australopithecus* cranium from Olduvai Gorge Bed 1 (Tobias 1967). The zygomatic arches are more strongly built and wider altogether than OH 5, but this is mainly illusory, since the missing incisor region and the palate, when reasonably reconstructed, have an almost identical shape in both. The cranial capacity has not been calculated directly, but is likely to be a little over 500 cm^3. The specimen is from area 10 in the lower part of the Upper Member of the Koobi Fora Formation.

KNM-ER 407 (Leakey 1971; Day et al. 1976)
This is a cranium missing the facial skeleton including the supraorbital tori. The cranium is lightly built and globular, with marked postorbital constriction and well-developed *M. temporalis* gutters in the zygomatic processes of the temporals. The temporal lines can be followed and come as close to each other as 21 mm. The greatest width of the cranium is at the supramastoid crests, and the posterior profile is bell-shaped. The mastoids are well developed and extensively pneumatized. Details of some parts of the cranium are excellent, but crushing during fossilization had displaced many fragments that could not be returned to their original position. The cranial capacity is unknown, but likely to be close to

500 cm^3. The specimen comes from area 10 in the lower part of the Upper Member of the Koobi Fora Formation.

KNM-ER 732 (Leakey 1971; Leakey et al. 1972)
This is the partial cranium of an adult, consisting of most of the right and parts of the left facial skeleton, together with the frontoparietal part of the calvaria and the right temporal bone. Only half the second premolar crown remains together with the roots (or alveoli) of this tooth through the third molar and a single incisor. The midline can be found, and the temporal lines do not meet in a crest but come close to the midline just posterior to bregma. The zygomatic processes of the temporal are very large and guttered superiorly for the temporalis muscle. The cranium is globular and thinly built. The supraorbital tori are not strong and the postorbital constriction is great. The face is flat, with widely flaring zygomatic processes of the maxillae, and would have had a slight amount of subnasal alveolar prognathism (from reconstruction). The roots and alveoli are large and the half premolar is 9.0 mm mesiodistally. The mastoid is large and well pneumatized. The specimen comes from the lower part of the Upper Member of the Koobi Fora Formation in area 10.

KNM-ER 733 (Leakey 1971; Leakey and Walker 1973)
Although a fragmentary specimen, this partial skull has preserved the right mandibular body and third molar, part of the right parietal, much of the left maxilla, most of the left frontal, most of the right zygomatic, and many parts of the cranial vault. From the teeth and the alveoli in the maxilla, this specimen is one like KNM-ER 406, with large posterior teeth and smaller anterior teeth. The conformation of the frontal is also very like that of KNM-ER 406, with the frontal rising very slowly, and with no strong curvatures from the moderately sized supraorbital tori. The zygomatic is massive, and, if oriented to give a realistic orbital margin, was extremely wide and flaring. The temporal lines, though raised, do not meet in a crest on the preserved parts. The cranial capacity cannot be accurately estimated. The specimen comes from within the Okote tuff complex (Lower/Middle tuff complex of Ileret) of the upper part of the Upper Member of the Koobi Fora Formation in area 8.

KNM-ER 1470 (Leakey 1973b; Day et al. 1975)
This is a partial adult cranium, missing most of the base, and parts of the facial skeleton. There are no tooth crowns, but the matrix-filled alveoli of the anterior teeth and the broken roots and alveoli of the rest of the teeth are present. The cranial capacity is between 770 and 775 cm^3, and the vault is thin and lightly built. The supraorbital tori are not salient. The greatest width of the cranium is

at the supramastoid crest and the posterior profile is bell-shaped. Postorbital constriction is moderate. The facial skeleton is large, being some 95 mm from nasion to alveolare. The face is fairly flat, with the upper edges of the pyriform aperture everted. The alveoli and roots suggest large incisors and canines and moderately large posterior teeth. There is some distortion in the specimen. The midline relationships are not affected, but plastic deformation and/or rotation of some elements have made the left and right positions of porion very asymmetrical. Since porion position has been used to distinguish *Australopithecus* and *Homo* (Wood 1976), this is of some consequence (Walker 1976). Porion is relatively far posterior in *Australopithecus*. Figure 13.4 is an equal-angle stereographic projection of KNM-ER 1470 that shows the midline structures in correct alignment. The coronal and squamous sutures have been plotted and the coronal is as nearly symmetrical as might be expected in an undistorted skull. The squamous suture has, however, been depressed on the right side so that the right temporal and porion have been carried downwards and forwards. The opposite movement, that of the left temporal moving upwards and backwards, can be rejected easily, since there is no overlapping or crushing at the squamous suture or elsewhere. On the right side, the squamous suture has slid open and there is a gap in the right side of the cranium. The forward movements of part of the right temporal can be seen in the plotted positions of other parts of it. Hence the position of porion in this cranium is close to the mean of *Australopithecus*. The specimen was collected from the upper part of the Lower Member of the Koobi Fora Formation in area 131.

KNM-ER 1590 (Leakey 1973a; Day et al. 1976)
This is a partial cranium of a juvenile with only the frontal and parietal bones, petrous, and some fragments of the calvaria. Several teeth and developing tooth crowns are preserved. The cranium was evidently large, since direct comparison with KNM-ER 1470 suggests that 1590 was bigger, even though it still had its second deciduous molar and deciduous canine in place and the central incisor was not quite in occlusion. The developing crowns of the anterior teeth are very large indeed, and those of the posterior teeth are also moderately large. The specimen is from area 12 in the upper part of the Lower Member of the Koobi Fora Formation.

KNM-ER 1805 (Leakey 1974; Day et al. 1976)
This is a nearly complete skull, missing only the rami of the mandible, most of the lower teeth and the supraorbital ridges and zygomatic bones. The elongated, ovoid braincase is small (cranial capacity about 580 cm^3) and has massive nuchal crests and a tiny, bifid sagittal crest well back on the parietal bones. The post-

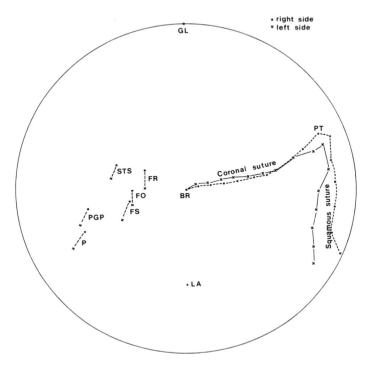

FIGURE 13.4
Sterographic projection (equal angular) of features of KNM-ER 1470 in *Norma verticalis*. See Oyen and Walker (1977) for details of method. The coronal sutures are aligned, but the right temporal has been rotated, opening the squamous suture. Left and right sides have been plotted on one side for comparison. Points on the basal side of the temporals have been plotted on the other side of the plot. In all cases the right side has been moved forward. Abbreviations: GL, glabella; BR, bregma; LA, lambda; PT, pterion; STS, sphenotemporal suture; FR, foramen rotundum; FO, formen ovale; FS, foramen spinosum; PGP, postglenoid process; P, porion.

orbital constriction is moderate. The facial skeleton has been crushed, so that the teeth and roots have been displaced in their alveoli and pieces of maxillary and palatal bone moved relative to each other. My colleague, Linda Perez, has undertaken the reconstruction of the facial fragment. This has been done, not on the original, but on a carefully painted plaster cast, using the original as a control. The various displaced parts were repositioned according to their edges and curvatures. The before and after photographs from this exercise are shown in Figure 13.5. The original palate was far too broad to occlude properly with its mandible. Ms Perez has aligned the maxillary fragments using the undistorted mandible as a guide. The resulting arcade is a smooth parabolic one. The teeth are of moderate size. The face is broad in the nasal region. The facial fragment can be articulated using anatomical features of the orbit as a guide, and the result is shown in Figure 13.6. The specimen comes from area 130 in the lower part of the Upper Member of the Koobi Fora Formation.

KNM-ER 1813 (Leakey 1974; Day et al. 1976)
This is a nearly complete cranium, missing part of the left facial skeleton, some of the base, and part of the occipital. Nearly all teeth are represented, on either the left or right. The elongated ovoid cranium is small, with globular braincase of about 510 cm^3 in capacity. The vault is thin, and in posterior profile has a bell shape. The widest part of the cranium is at the supramastoid crest. The postorbital constriction is moderate and the supraorbital tori are not particularly strong. The facial skeleton is large relative to the braincase. Despite the fact that the teeth are only of modest proportions for an early hominid, they are relatively large. The temporal lines do not come close to the midline. The gutters for the posterior fibers of *M. temporalis* are only moderately developed. The most striking resemblance is between this specimen and Olduvai Hominid 13 (Leakey and Leakey 1964; Tobias and von Koenigswald 1964). In practically every detail of size and shape of the teeth and all preserved parts, these two specimens are extremely close. Figure 13.7 shows the palate of both specimens. This is of great interest, because of all the specimens assigned to *Homo habilis*, Olduvai Hominid 13 is the only one that most authorities agree was in, or close to, the ancestry of *Homo erectus*. In fact, it has been reconstructed with an *erectus*-like vault (Brace 1973). Now we can see from comparison with 1813 that Olduvai Hominid 13 was probably not much like *Homo erectus*. Rather, it probably had a facial skeleton, palate, and teeth of the size of an *erectus* set on a very small calvaria. Even with a mandible containing all its teeth, and much of the rest of the cranium, it seems that this is not enough to accurately determine the affinities of a particular hominid. The specimen comes from area 123 and its stratigraphic position is still not certainly known. Provisionally it is placed in the lower part of the Upper Member of the Koobi Fora Formation.

KNM-ER 3733 (Leakey and Walker 1976; Leakey, Rose, and Walker 1978)
This is a nearly complete cranium, lacking parts of the lower facial skeleton and certain teeth. The cranium is large (cranial capacity about 850 cm^3) and very thick. The lateral profile is long and flattened, with a pronounced occipital

206 Homo erectus

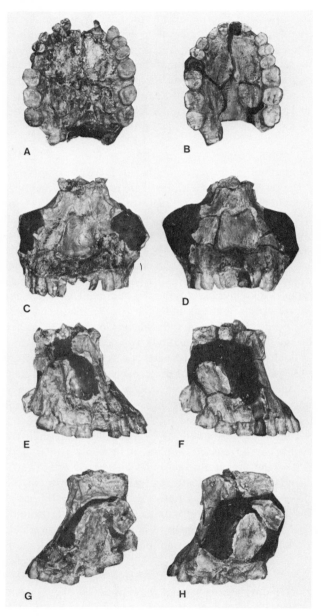

FIGURE 13.5
Specimen KNM-ER 1805 before and after reconstruction.

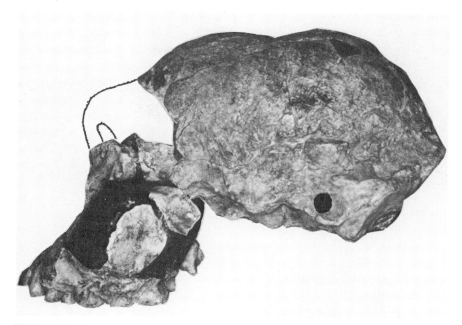

FIGURE 13.6
Lateral view of reconstruction of KNM-ER 1805 cranium.

projection and has a large part of the braincase behind the external auditory meatus. The supraorbital tori are projecting, but not particularly massive. There is a distinct postglabellar sulcus. and the frontal rises steeply behind it to reach vertex at bregma. The facial skeleton is relatively small and tucked well under the braincase. The zygomatic portions are deep. The nasals are longitudinally concave and laterally convex and project over a wide, low pyriform aperture. The palate is roughly square in outline, with the incisor alveoli set in almost a straight line. The anterior teeth are moderately large, as judged by their alveoli, and the posterior teeth are of only moderate proportions. Although both third molars are missing, they were present in life, because the distal faces of the second molars bear contact facets.

No traces of third molar roots remain, so the chances are high that the crowns and roots were reduced in size. The specimen is very reminiscent of those from Peking (Weidenreich 1943) but it is more complete than any of the Chinese specimens. The cranium comes from area 104 and was in situ in the lower part of the Upper Member of the Koobi Fora Formation.

208 Homo erectus

FIGURE 13.7
Comparative palatal views of KNM-ER 1813 and Olduvai Hominid 13.

KNM-ER 3732 (Leakey 1976a)
This is a partial cranium that has the supraorbital tori and much of the top of the cranial vault preserved, together with part of the left facial skeleton. In all preserved parts it closely resembles KNM-ER 1470. It comes from the upper part of the Lower Member of the Koobi Fora Formation in area 105.

KNM-ER 3883 (Walker and Leakey 1978; Leakey, Rose, and Walker 1978)
This is a cranium that lacks the facial skeleton, except for parts around the nasals, the right orbit and right side of the pyriform aperture. In nearly all respects it is very similar to KNM-ER 3733. The supraorbital tori are a little more massive and, as a consequence, the postglabellar sulcus is shallow. The mastoid and preserved parts of the facial skeleton are also more robust than in KNM-ER 3733. The cranial capacity has not yet been recorded, but it is likely to be close to 850 cm^3. This specimen comes from the lower part of the Upper Member of the Koobi Fora Formation in area 3.

KNM-ER 3884 (Walker and Leakey 1978)
This is a crushed cranium, with all its teeth, from the Guomde Formation. It is of great interest because of its age – about 700,000 years. Preparation has begun, but no statements can be made at present about taxonomic affinities. This specimen comes from area 5.

This list includes all specimens that offer more than just fragmentary information about the cranium. What sense can be made of this sample? Taking first the sample from the Upper Member of the Koobi Fora Formation, that is, those dated between about 1.2 million years to either about 1.6 million years or about 2.4 million years, it appears that there are three forms present. These are: (1) hyper-robust crania with large facial skeletons and postcanine teeth, and small braincases; (2) crania with small braincases and smaller teeth and faces, that nonetheless still have large faces and teeth relative to the calvaria; (3) crania that show remarkable similarities to those from Java and China which have been called *Homo erectus*. These have larger, low braincases and relatively small facial skeletons and teeth.

Rather than trying to assign these specimens to known taxa, it is more instructive to see what are the possibilities of taxonomic arrangement. Figure 13.8 outlines the proposed scheme. There are 5 possibilities that can be considered:

1 They are all from one species, and the finding of three forms is just an artifact of sampling or pattern recognition. That is, in fact, the single species hypothesis (Brace 1967; Wolpoff 1971a), which states that there has only ever been one hominid species. The idea is based on a series of assumptions about the adaptive nature of hominid morphology and on the principle of competitive exclusion (Gauss 1934). Bluntly put, adherents believe that the basic hominid characteristics of bipedal locomotion, reduced canines, delayed physical maturity, and so on have come about in response to more effective and greater dependence upon culture. The assumption of a basic hominid cultural adaptation, taken together with Gauss's principle, leads to the conclusion that two hominid species would not be able to exist sympatrically.

2 They are all separate species. In this case, the hyper-robust crania would be placed in the taxon *Australopithecus* (or *Paranthropus*) *robustus* (or *boisei*). They have their greatest similarities with specimens from Olduvai (Olduvai Hominid 5), Chesowanja (Carney et al. 1971), Swartkrans (Broom and Robinson 1952), and Kromdraai (Broom and Schepers 1946). The fact that, on the whole, the East African specimens are larger could be explained by placing them in a separate species (*A. boisei*), or by claiming that they represent northern populations of *A. robustus* and that a geographic size cline exists. Faunal indications of ages of the South African sites would rule out the possibility of one group being ancestral to the other, since all of them are essentially contemporaneous (White and Harris 1977).

Possibilities 3, 4, and 5 group pairs of forms together within one species, leaving the other form as a separate species.

210 *Homo erectus*

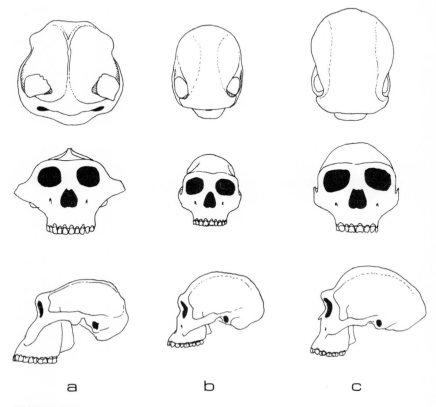

FIGURE 13.8
The three forms of crania found in the Upper Member of the Koobi Fora Formation, and the five possibilities of their groupings.

3 The robust crania and the small-brained, small-toothed forms are from one species and the *Homo erectus* are separate. Here the smaller individuals would be females of a sexually dimorphic species. This would mean accepting a normal amount of cranial capacity dimorphism, but an unusual amount of dimorphism in tooth proportions and size.

4 This possibility is that the robust crania and the large-brained crania would be from the same highly variable species, and that the small-toothed, small crania are in one, less variable species. This second species would probably be *Australopithecus africanus* and would have its affinities with specimens from Sterkfontein (Broom and Robinson 1950). In order to accept the other crania in a single

species, we would have to accomodate both extremes of braincase variability and dental variability.

5 The last possibility is that the robust crania are in a separate species and the small-toothed forms are in their own, highly variable species. In this last case, the amount of dental variability is small, but the cranial variability is very great, not only in size, but also in shape.

Having decided that there are only these five possibilities (if we recognize the three forms) it remains to assign the probabilities that each case is correct. These need not be, at this stage, actual numerical probabilities. For the first possibility, it is my opinion that the probability of it being correct is close to zero. The demonstration of two fine *H. erectus* crania and robust *Australopithecus* in the same time period amounts to a rejection of the single species hypothesis. To place all these fossils in the same species would be to assign to past populations much greater variability than can be observed in any living hominid ones. It seems to me that the masticatory adaptations in the robust *Australopithecus* crania overshadow the braincase in a quite opposite way to the overshadowing of the dentition and facial skeletons by braincase seen in members of the genus *Homo*.

Possibility 2 has, in my opinion, the greatest chance of being correct, despite the fact that some specimens, especially those that are fragmentary, might be incorrectly placed. For instance, it is possible that the small crania, KNM-ER 732 and KNM-ER 407, could be female *A. robustus* as Leakey (1971) proposed (see also Robinson 1972b). This would mean, however, that even more fragmentary specimens than these would be impossible to sort into types and thus, even if it has occurred, this error would be difficult to substantiate. The specimen, Sts 71, from Sterkfontein shows a number of features that are reminiscent of these two East Turkana crania, among them the wide zygomatic process of the temporal, the fairly flat frontal region, the mastoid inflation and the large zygomatic process of the maxilla.

Possibility 3 has a lesser chance of being correct, but the possibility must be faced that dimorphism within *Australopithecus* might have been greater than anything encountered today. The cranial capacity variability would be within the bounds of present-day values for humans and pongids. The dental variability would be greater than encountered today.

Possibility 4 has very little chance of being correct, for the very same reasons as number one. If skulls as different as robust *Australopithecus* and *Homo erectus* are from the same African populations, then where are the robust specimens at Peking and in Java? The Meganthropus specimens, so often brought forward as being Asian robust *Australopithecus*, are not convincing (see Tobias and von Koenigswald 1964).

The fifth possibility suffers from the same problem as the fourth. That is, although dentally the specimens could be accounted for in one very variable population, the cranial capacity dimorphism seems far too extreme. Besides this, the cranial construction of the *H. erectus* crania and the all-important features of the relationship between facial skeleton and braincase is clearly very different from that of the small-brained, small-toothed form. The same question can be asked here about the Asian sites. Where are the crania that look like KNM-ER 1813 in China and Java?

To summarize, it seems that the greatest probability is that there were three contemporary species during this time period. Taxonomically, these would be *Australopithecus boisei* (or *robustus*), *Australopithecus* cf. *africanus*, and *Homo erectus*. There is a lesser probability that the first two are really from one taxon, and there is hardly any probability that the other three possibilities are correct.

During the earlier time period, that is from the upper part of the Lower Member of the Koobi Fora Formation, the cranial sample is not large. The three crania that have been found below the KBS, or its lateral equivalents, are all of the form exemplified by KNM-ER 1470. They all have indications of a cranial capacity between 700 and 800 cm^3. They all show an *Australopithecus*-like construction of the cranial vault. The facial skeleton of one, together with the teeth of another, show that the facial skeleton was extremely large relative to the braincase, as in *Australopithecus*. There is mandibular and dental evidence that at least the small *Australopithecus* was also present. It is the taxonomic status of these crania from below the upper part of the Lower Member that is important in understanding the origins of the genus *Homo*.

Some time ago I had the task of writing about the specimens belonging to *Australopithecus* from East Turkana (Walker 1976). I found that before I could do so I had to consider the factors that should be thought of when making a generic diagnosis. The points considered ranged from the trivial to the difficult, but my colleague Eric Delson chastized me gently for not paying enough attention to cladistic analysis. I have read those efforts that have been published concerning hominid evolution and have found myself confused (and sometimes dismayed), as have Kay and Cartmill (1977), at the inconsistencies in what is supposed to be a highly consistent exercise. At the heart of the trouble, I suspect, lie the problems of how to deal with great variability in a tiny sample, and the problems of deciding upon primitive characters. I will admit, however, that the recent rush of claims for early specimens of the genus *Homo* from both East and South Africa most likely stem from workers emphasizing primitive states. Those of the school of cladistics are, I feel, also deluding themselves that any particular phyletic relationship may be testable in any normal sense of testability (see Kuhn 1970, for instance, for a discussion of Karl Popper's views).

The only realistic way of dealing with the present tiny sample in a cladistic way that avoids the problems of variability, might be to deal with the single specimens, imperfect as they are, and construct cladograms for them. This would remove some difficulties, but the record is still very inadequate. I find the cladistic attempts to date excessively typological, yet I know how difficult it is to avoid thinking of, say, Sts 5 when dealing with Sterkfontein, since good specimens are so few. In fact, there are *no* complete crania or skulls of any early hominids. As for dealing with postcranial remains, we must continually remind ourselves that we have *no* specimens of any early hominid individual in which decent cranial and postcranial remains are associated. The Afar skeleton (Johanson and Taieb 1976) is the best by far, but there is not much of the skull in that magnificent specimen. There are *no* associated cranial and postcranial remains, from any site, of robust *Australopithecus*.

I find, on balance, that the crania from the Lower Member are best placed in the genus *Australopithecus*, their major differences from *A. africanus* being probably only of size. Nevertheless, I think that they represent the antecedent population to the *H. erectus* population that is sampled in the Upper Member, for to postulate an Asian or Eurasian origin for the latter population is unnecessarily complicated. If indeed *H. erectus* evolved outside of Africa, it entered just when one part of the *Australopithecus* radiation had gained larger sizes (including brain size), and had already begun to litter the East African landscape with stone tools.

The species for these specimens would probably be *Homo habilis* (Leakey et al. 1964), but not in their sense entirely. Only Olduvai Hominids 7 and 16 fit into this category in my scheme, and since both are so very fragmentary, it is difficult to be dogmatic. Other specimens assigned to *H. habilis*, such as Olduvai Hominids 13 and 24 are, to my mind, part of the small-toothed small-brained species that I regard as an East African, late population of *A. africanus*.

Several other points are brought out by this assessment. The composite cranium, SK 847, from Swartkrans (Brain 1970) that has been an advanced hominid for some and an *Australopithecus* for others, seems to be a very fragmentary specimen of a very old adult *H. erectus* (see Figure 13.9). The similarities to KNM-ER 3733 are striking, and I am no longer convinced that the midline relationships are any guide as to the real-life ones, especially since the specimen has suffered some postfossilization damage. As it happens, the faunal relationships show that SK 847 and KNM-ER 3733 are of approximately the same age (White and Harris 1977). Apart from showing that *H. erectus* was contemporary with a large robust *Australopithecus* at Swartkrans, as it was at East Turkana, this also shows that it is hardly worthwhile dealing with any fragmentary specimens if a new taxon is being recorded. As hominid specimens

214 *Homo erectus*

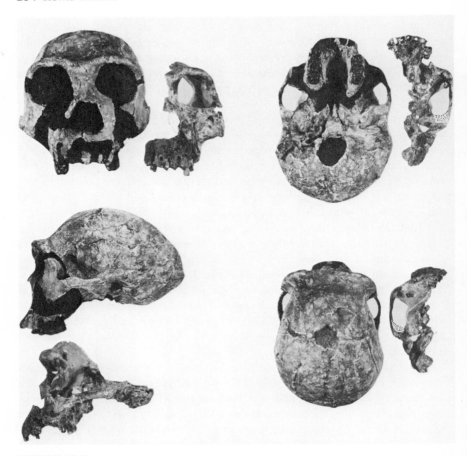

FIGURE 13.9
Comparison of the more complete cranium KNM-ER 3733 and the composite cranium SK 847 from Swartkrans.

go, SK 847 is a reasonably complete one, yet it is not complete enough to satisfy skeptics. It takes, apparently, a specimen as good as 3733 for that, and 3733 is one of the most complete fossil hominid crania known.

The East Turkana fossils, although only a tiny sample of past populations, show us that the hominids underwent a minor radiation about two million years ago. One of the species, *H. habilis*, was a larger version of the species ancestral to all of them. This large species underwent changes that transformed it to *Homo erectus*. The time period required for the changes was either very short (on the order of about a few hundred thousand years), or quite long (about a million years). Resolution of the problem of the age of the KBS tuff will tell us whether our own genus arose rather suddenly or gradually (Eldredge and Gould 1972). This should be one of the most urgent tasks for our colleagues in the relevant disciplines. At a time when modes of evolution are beginning to be discussed in detail, the matter of our own origins should be of the highest priority.

14

REINER PROTSCH

The Kohl-Larsen Eyasi and Garusi hominid finds in Tanzania and their relation to *Homo erectus*

Cet article s'occupe en premier lieu de la réexamination et de la réévaluation de deux hominidés fossils qui ont été trouvés lors de l'expédition de Kohl-Larsen entre 1934 et 1940 à Tanzania: les hominidés de Eyasi (*Africanthropus njarasensis*) et de Garusi (*Meganthropus africanus*). La reconstruction des hominidés de Eyasi comme *Homo erectus* par Weinert est mise en doute. D'après une nouvelle analyse morphologique et une datation absolue, ils appartiennent au Pléistocène supérieur àpeuprés 35,000 av. aujourd'hui, bien ultérieurement à l'espace de temps de *Homo erectus*. Selon la nouvelle analyse, ils appartiennent très probablement à *Homo sapiens rhodesiensis*.

Par contre, les hominidés de Garusi ont été trouvés dans les couches de Laetolil (un fragment maxillaire et un M³ séparé) et actuellement ils sont les premiers hominidés découverts à Laetolil. Ils appartiennent au genre *Homo* et très probablement remontent au moins à 3.5 millions années av. aujourd'hui.

Autrefois les hominidés de Garusi ont été considérés soit comme *Homo erectus* soit comme *Australopithecus*. Une nouvelle analyse morphologique, pourtant, assigne clairement les hominidés de Garusi aux hominidés de Laetolil, précedent ainsi *Homo erectus*.

Traditionally the Eyasi and Garusi hominids have been attributed to *Homo erectus*, although the Garusi hominid presents a more questionable placement in the literature. Hans Weinert, the main proponent of their taxonomic placement, quite probably worked more in accordance with preconception rather than with the usual scientific procedure. It is often stated that his work on this fossil material was meticulously exact and extensively carried out. However, even after a short examination of the original reconstruction as well as of additional pieces, it should become quite obvious to the trained morphologist that Weinert's reconstruction is questionable.

A new reconstruction of these finds has been attempted. In their present state a careful description of new details is possible that provides new clues to their taxonomic status (Protsch 1980). It is my intention to give a brief history of these finds including a short description of them, their state of preservation, number, and importance of additional pieces, and their assignment to specific individuals (i.e. Eyasi I, II, III, IV). Dates, or the possibility of dating, as well as suggestions about taxonomic grouping will also be touched upon. A general comment will be followed by specific sections dealing first with the Eyasi and then the Garusi hominids.

Compared to other geographical areas of the world, East Africa, and specifically the area around and close to Olduvai Gorge, has yielded only scarce material of hominids that could be assigned to or with the *Homo erectus* group. At a time when only the find of Olduvai Gorge known as Hominid 1 (OH 1), which was discovered in 1913, and a few somewhat more recently dated hominid finds (ca up to 15,000 years BP) were known, two finds aroused considerable interest among paleoanthropologists. These finds were those from Eyasi and Garusi, discovered by the German Kohl-Larsen Expedition of East Africa during several excavation seasons from 1934 to 1940 (Kohl-Larsen 1943; Reck and Kohl-Larsen 1936).

Today new finds from this area, some with earlier dates, are readily available to anthropologists either in the form of casts or original material. In the past, however, only a few specialists had access to such originals, and the production of casts for other specialists was generally neglected. Under these circumstances it is not surprising that some mistakes were made by the few anthropologists who studied these finds — mistakes that could be corrected only many years later. Photographs could give only limited information about the morphology of these finds, and the restudy of some originals was hindered by their removal from institutions for safekeeping during World War II, a circumstance which, as in the case of the Eyasi hominids, led to their disappearance for some years thereafter.

After the Eyasi originals were rediscovered, studies of their morphology in the early 1970s included a new reconstruction and the application of new analytical methods like chemical-physical types of dating. This served to give us a more complete and correct picture of these hominids.

In the late 1930s and early 1940s, after only short and preliminary morphological investigation, the Garusi and Eyasi fossils were too readily assigned to a group of fossil hominids that were not yet known to exist in East Africa or for that matter on the whole African continent. Their study by H. Weinert and others was only seemingly extensive (Weinert 1937, 1938a, 1938b, 1938c, 1940, 1952). Weinert's hasty assignment of these finds to the *Homo erectus* and pre-*Homo erectus* groups, even before their reconstruction, should leave some doubt as to the correctness of their proposed position in hominid phylogeny. The Eyasi hominids consisted of so many fragmentary pieces (over 250) that only a very diligent and lengthy reconstruction could solve the question of their taxonomic and chronological position. In the case of Garusi, only a small part of the right maxilla with first and second premolars in place, as well as a separate molar (upper third), was found with the alveoli of the canines and incisor well preserved but still covered with matrix until 1973. Previous scholarly evaluations were based on only a few poorly made casts that not only gave a distorted picture but also hindered exact morphological evaluation.

With the availability of new analytical methods and the retrieval of additional fossil bone material it has been possible to better reconstruct and evaluate both hominids. The methods included the use of more delicate cleaning equipment and absolute dating techniques enabling precise chronological placement.

To be more explicit, the following points now make it possible to assign both hominids positively into more specific fossil hominid groups: (1) new morphological reconstruction; (2) removal of matrix covering important parts of the fossils; (3) relative as well as absolute placement in time; (4) morphological comparison to fossils of the same locality and stratigraphy; (5) mineralogical evaluation of the matrix adhering to these fossils and its comparison to new hominid finds with the same morphology and of the same stratigraphy; and (6) application of indirect A_1-dating by potassium-argon of the new Laetolil hominids to some of above fossil hominids (A_3-date).

THE EYASI HOMINIDS

Contrary to accounts in most text books, the Eyasi hominids were not 'lost' during World War II but stored on a farm for safekeeping. They were relocated at Tübingen, West Germany in the late 1960s. Located were all hominids presently known as Eyasi I, II, III, and IV. Eyasi II, an occipital fragment, was returned to Dar-es-Salaam, Tanzania, several years thereafter. All of the fragments, including some associated faunal material, were originally found during two excavation seasons from 1934 to 1940. The reason why a German expedition was still doing research in this area is that it was a former German colony (German East Africa). This meant that the researchers had detailed knowledge of the region through a series of expeditions by the military, by geologists, and by a number of other scientists during the period from 1876 to 1914. L.S.B. Leakey had not yet started his extensive study of the area.

Lake Eyasi is situated 50–55 km south of Olduvai Gorge, and the area of the fossil finds is located in a bay at the northeastern part of the lake. The bay itself is erroneously mentioned in the Kohl-Larsen expedition report as the western bay. The excavation of the locality was initially to be a small project. The major goal of the expedition was the collection of ethnological material on the Tindiga tribe inhabiting the area around Lake Eyasi (then called Lake Njarasa by the Germans).

All excavated materials were surface-to-subsurface finds contained in a sandstone layer with a depth of about 35 cm. Hominids, fauna, and artifacts were found randomly mixed together in this layer, which is located directly on the lake shore. Although a more extensive excavation was planned because of the promising nature of the initial finds, it was not carried through because of problems with groundwater and the migration of surface waters, which frequently came to within 30-100 m of the site. Thus an intensive search for additional material was considerably hindered. The major hominid find, Eyasi I, was discovered on November 29, 1935. It was assigned a relative chronological age on the basis of associated fauna and artifacts.

This quite fragmentary material, found scattered over an extensive area, numbered over 200 pieces. They were used in a preliminary reconstruction by H. Reck and L.S.B. Leakey in 1936. Although Leakey assigned Eyasi I, according to its primitive characteristics, to a fairly early time period with a possible close relationship to the *Pithecanthropus* group, he cautioned against using solely this criterion (Leakey 1936, 1947, 1948, 1952). He believed that associated finds had to be considered and more detailed reconstruction had to be attempted. In light of Leakey's cautious consideration of the facts it is especially surprising that Weinert (1937), after only a short examination, unhesitatingly gave the fossil the name *Africanthropus njarasensis*, a *nomen nudem* already assigned to the Florisbad hominid. He positively aligned it with *Pithecanthropus*. In doing so Weinert not only disregarded Leakey's, Reck's, and Kohl-Larsen's thorough investigations of geological deposits in which the fossils were found, but at the same time tried to defend himself against criticism by F. Birkner (1937). Birkner saw Eyasi as a possible member of the Neanderthals in Africa. When Bauermeister, Weinert's assistant, attempted a new reconstruction in 1939 (Weinert et al. 1940) he had changed little from the original one by Leakey and Reck. Only a few additions and alterations were made on the frontal and occipital bones. These parts were, however, among the most decisive in assigning the fossil to either *Homo erectus* or to the African Neanderthals. To be more specific, very few of the fragments, except three in the region of the parietal bones, fit together in Weinert's reconstruction. Most pieces are separated by large gaps giving the entire original reconstruction a somewhat disjointed and awkward appearance.

The resultant morphological picture resembled *Homo erectus* in many aspects. Numerous small pieces found separately couldn't possibly be included in Weinert's reconstruction in that state. Weinert pointed out (1938d) that the hominid remains were remnants of cannibalism. He based his assumption on the quite fragmentary nature of the bones. There is no evidence that would support such a theory concerning the Eyasi hominids. After microanalysis (F, U, N), these fragments could easily be assigned specifically to Eyasi Hominid I, II, III, or IV. This relative placement technique was not available to the original researchers of the 1930s and early 1940s (Protsch 1976), the value of which in this case is quite critical. For example, because some pieces were found as far away as 430 meters from the original site of the 1934 excavation, they had been immediately assigned to a hominid other than Eyasi I. All of these are smaller fragments with more diploë than outer or inner table and were thus of light weight. This indicates that they were probably transported by water and wind in a southeasterly direction and then redeposited. With microanalysis it is possible to assign the many separate pieces to individual hominids. Indeed, most of the pieces belong specifically to Eyasi I, according to microanalysis. Thus we were able to add additional pieces to Eyasi I resulting in a new, more complete reconstruction.

Palaeoanthropus njarasensis, the second name given to the Eyasi hominids, seems to belong to the 'African Neanderthal' group, *Homo sapiens rhodesiensis*, rather than to *Homo erectus*. Absolute dates assigned to Eyasi I by amino-acid dating run by two different laboratories designate dates of 34,000 and 35,000 years BP (aspartic acid) (Protsch 1975). These dates support the assignment of the hominid to *Homo sapiens rhodesiensis* and allow proper taxonomic evaluation on the basis of the new morphological reconstruction. To strengthen this argument it should also be pointed out that there currently is no paleoanthropological evidence to support the existence of *Homo erectus* in this late time period.

Eyasi I, and possibly the other Eyasi hominids, can thus be added to the other members of *Homo sapiens rhodesiensis*, which include Saldanha, dating to ca 40,000 years BP (Protsch 1975), and Broken Hill, possibly dating to ca 110,000 years BP (Bada et al. 1974). The Broken Hill date, also an amino-acid date, should be used with caution and viewed with skepticism. As in all cases this technique should be used in conjunction with other supportive absolute dating techniques. Amino-acid dating, which can be affected by temperature and water, might also be influenced by associated metals, as is the case with Broken Hill.

The morphology of Eyasi I is not too different from Saldanha and Broken Hill but quite dissimilar from *Homo erectus*, assuming there actually exists a 'typical' *Homo erectus* morphology. According to initial estimates the cranial capacity of Eyasi I is about 1220 to 1250 cm^3, a measurement close to that of the Broken Hill individuals. The skull curvature seems also to align this hominid with Broken Hill, whereas the occipital does not differ substantially from the late Middle Pleistocene hominid of Steinheim (also Weinert 1940). An alignment according to anatomical normae shows that in Weinert's reconstruction *Norma occipitalis*, *Norma frontalis*, *Norma parietalis*, and *Norma verticalis* cannot be correct. It is impossible to see a continuous occipital torus, a supraorbital torus, or a tight postorbital constriction, features which are said to be typical of *Homo erectus*. The skull is not as platycephalic and the nuchal crest is not at all horizontal as is characteristic of *Homo erectus*. The maximum skull breadth does not seem to occur toward the lower part of the skull, on the temporals, but is instead high on the parietals. The thickness of the cranial bones is normal and falls into the range of anatomically modern man. The above observations support exclusion of the specimen from the *Homo erectus* group.

The internal morphology (completely neglected by Weinert) gives, after removal of all plaster parts, valuable additional information that can be used to refit pieces properly. In many cases the inner table is well preserved, contrary to Weinert's assertion (Weinert et al. 1940), and can be used for proper fitting. Besides this, over fifteen additional pieces can now be utilized in the new reconstruction. This allows a more proper and accurate fitting of sutures and pieces such as a part of the left maxilla with a canine and first premolar belonging to Eyasi I.

Many pieces of the former Eyasi II and Eyasi III are faunal material; other pieces are very fragmentary hominid material, with only a few pieces preserved well enough to recognize morphological features justifying an inclusion in *Homo sapiens rhodesiensis* (Protsch 1980).

On the basis of morphology the Eyasi individuals should be separated from other *Homo erectus* individuals of East Africa. The Eyasi hominids are neither similar to Olduvai Hominid 9 (OH 9) nor, judging from published photographs, to the new Ndutu specimen assigned to *Homo erectus*, a claim made by Clarke (1976). It is advisable that more detailed analysis and comparison should be attempted in the near future.

The Eyasi hominids are not the only hominids excavated since 1940 around Lake Eyasi. In 1967 a short excavation season was conducted by J. Ikeda of Kyoto University, Japan, at Bangani, Lake Eyasi. Although no detailed morphological descriptions have yet been published, Ikeda believes that the two Bangani mandibles excavated are contemporary with the Eyasi I hominid (Ikeda, personal communication 1975).

On geological, paleontological, and cultural grounds as well as on the basis of relative and absolute chronological comparisons, it seems impossible that the Eyasi hominids could date anywhere between 600,000 and 500,000 years BP as seems to be the case with the Ndutu individuals (Mturi 1976).

In conclusion, my findings support the original interpretation of Leakey, Reck, and Kohl-Larsen based on geological, cultural, and faunal evidence for an Upper Pleistocene age of around 35,000 years BP (Reeve 1946).

THE GARUSI (NGARUSI) HOMINID

This hominid was found in February 1939, on the northwest side of Lake Eyasi inside a gorge of the Garusi River. The location is closer to Olduvai Gorge than that where the Eyasi hominids were found. The excavation was rather short because war was imminent between England and Germany and the German expedition was under constant fear of interruption by British authorities. Contrary to reports in the literature, more material than the famous right maxillary fragment with first and second premolars, commonly known as *Meganthropus africanus*, was found. On February 20, other skull fragments were found. Among them was a large occipital fragment that was subsequently lost in Germany. The state of fossilization of this fragment was somewhat different from that of the maxillary fragment found previously, thereby negating the possibility of close affinities between the two.

On February 24, 1939, a separate molar (right M^3) and another occipital fragment were found. As the second occipital fragment displayed another unique state of fossilization, it too was discarded and only the molar was kept. These finds came from the same tuffaceous deposit now known as the Laetolil Beds, situated close to the Deturi (Laetolil) River. The separate molar was first thought to be part of the same individual to which the maxillary fragment belonged. Later investigators, however, excluded the molar from the maxillary specimen known as *Meganthropus africanus* because the occlusal surface showed completely different attrition from the premolars of the maxillary fragment. The separate molar has three roots and its overall size suggests that it could have belonged to the same maxilla holding the two premolars. Its length is 10.0 mm and its breadth is 13.8 mm. These measurements are slightly higher than those of modern Australians and Bantus but below those of the australopithecines of Swartkrans and Sterkfontein. They would be similar to *Pithecanthropus* IV.

Microanalysis of bone adhering to the roots of the molar, as well as its state of fossilization, and the mineralogical makeup of the tuff deposit within the molar and maxilla suggest a contemporaneity of these separately found items. Although both finds date to the same horizon and time period, they can, however, be assigned to two different hominids. Garusi I, the maxillary fragment with the two premolars, belongs to an individual of fairly young age, possibly between 20 and 25 years. Garusi II, the separate molar, undoubtedly belongs to a different and much older individual as the occlusal attrition is quite advanced. There is no possibility that such differentially worn teeth could belong to the same individual.

Weinert (1950) named the hominid *Meganthropus africanus* because he thought it was morphologically close to the *Meganthropus* of Java. This is a view not shared by most other specialists. Opinions range from assignation to *Sinan-*

thropus (Remane 1951, 1954), to a preanthropine stage (Hennig 1948; Senyürek 1955), to *Plesianthropus transvaalensis* and finally the australopithecines (Robinson 1953). Most specialists agree that the fragment is hominid, even though most had the ill fortune of working with poor casts. To add to the problem, the alveoli of the canine and lateral and central incisors were covered by a tuffaceous matrix. Microanalysis on F, U, and N of the molar and the maxillary fragment show the same results. Amino-acid dating was attempted but the fossils seem to be out of the range of this dating technique. Results of a previously run date, 800,000 years BP, are misleading because of contamination of recent amino-acids due to preservatives.

Cleaning of the alveoli of the canine and the incisors provided a better knowledge of the size and morphology of these teeth. When reconstructed it appears that the canine projected above the occlusal plane of the other teeth. It is quite large, substantially larger than canines of modern man (see Figure 14.1). Apart from that, the matrix adhering to the maxillary fragment was compared to the mineralogical makeup of the Laetolil Beds and has also been compared to the matrix adhering to the new Laetolil hominids found by M. Leakey during the 1973-76 excavations in the Garusi area (Leakey et al. 1976). Mineralogical analysis of the tuff adhering to both the Garusi and the new Laetolil hominids proves that they originate from the same level in the Laetolil Beds.

The number of roots of the premolars could finally be accurately determined by use of x-ray analysis. It was found that the first premolar had three roots, two buccal and one lingual, and that the second had two roots, one doubly fused buccal and one lingual root.

Although the teeth are larger, their morphology is not unlike that of modern *Homo sapiens*. The second premolar is, contrary to earlier reports, somewhat smaller than the first, an essentially modern trait. Because the first premolar has three roots it appears to be similar to some specimens from Swartkrans and to pongids. Mesial contact facets can clearly be recognized on both of the premolars, and on the third molar. They are all noticeable as concave areas on the upper part of the crown, which is an indication of some crowding of the postcanine teeth. The contact facet on the mesial side of the first premolar is located more buccally as is the case in most modern hominids. This again suggests that the contact zone of the canine is also more buccally located. It is, however, the small contact facet on the premolar that seems to indicate the presence of a fairly large canine. As mentioned above, a fairly large canine seems to have been present judging from the length of the alveolus as well as the breadth and length of the alveolar margin. The root itself is much larger than in *Australopithecus robustus* or *A. africanus*. Finally, strictly judging from the minimum thickness of the interalveolar septi, no diastema, either precanine or postcanine, seems to have been present. Even though only a small maxillary fragment is available, a

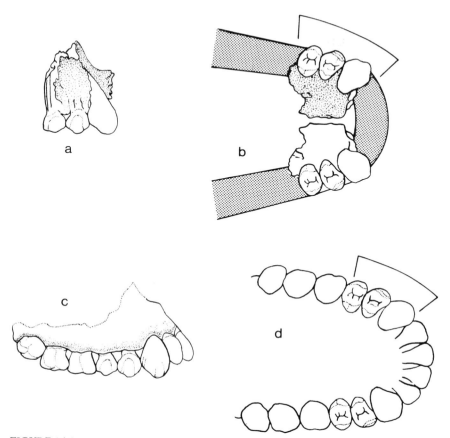

FIGURE 14.1
Garusi I (*Australopithecus sp.* or *Homo sp.*): a) labial view with reconstructed canine; b) occlusal view with reconstructed left maxillary section; c) lateral view of reconstructed total right maxillary section; d) occlusal view of totally reconstructed maxillary dentition.

reconstruction of the dental arcade is possible. The alveolar margin is curved convexly resembling that of modern man. The overall morphology and state of preservation of the maxillary fragment seems to be similar to the new Laetolil hominids (T.D. White, personal communication 1976). It is interesting to note that Senyürek as well as Oakley place the hominid into the Laetolil Beds (Oakley 1968).

A more detailed morphological examination is forthcoming in the Kohl-Larsen Festschrift (Protsch 1981), but it seems beyond doubt that the Garusi maxilla belongs to some member of the genus *Homo*. That it is a member of *Homo erectus* seems to be somewhat doubtful. Instead, it seems plausible to include the individual in a group leading to *Homo erectus*. The previous disagreement among scientists regarding its taxonomic position seems to strengthen this argument. Accordingly, the Garusi hominids can be assigned to the Laetolil Beds because of the mineralogical similarity in the tuffaceous matrix adhering to both Garusi and the Laetolil hominids (Leakey et al. 1976), and because of the similarity in morphology between the two sets of hominids.

CONCLUSION

It seems that neither the Eyasi nor the Garusi hominids are members of 'typical' *Homo erectus* in East Africa. It has been demonstrated both chronologically and morphologically that the Eyasi hominids belong to the taxon *Homo sapiens rhodesiensis*. The Garusi hominids, on the other hand, should be grouped with those found recently by M. Leakey at Laetolil, thus making them predecessors of *Homo erectus*.

JEROME S. CYBULSKI

Homo erectus: a synopsis, some new information, and a chronology

The purpose of this concluding chapter is to provide an overview of 'known' *Homo erectus* fossils. In addition to synthesizing material that has been presented in chapters 4 through 14, this section fills certain lacunae and summarizes information and views that have been published since those chapters were finalized and while the typescript of the volume was being prepared for the printer.

The intricacies of assigning fossils to the binomen, *Homo erectus*, and the validity of this taxonomic category, have been discussed by Howells in chapter 5; and we have witnessed the complexities of dealing with the issues of fossil assignment in nearly every following chapter. Indeed, one may well wonder whether agreement will ever be reached as to which fossils do belong to or represent the taxon, and on what morphological-cum-phylogenetic grounds fossil hominids are or are not to be regarded as *Homo erectus*. The following pages are devoted to an area synopsis of *Homo erectus* fossils and their temporal placements as currently viewed. Some other fossils relevant to these questions are also considered. Finally, an attempt is made to integrate the assemblage in a summary chart of time, space, and morphology.

CHINA

There appears to be virtual agreement that *Homo erectus* is well represented in China by the Locality 1 fossils of Chou Kou Tien (Peking Man), perhaps best delineated as no more than subspecifically distinct from *Homo erectus* of Java. The Locality 1 series is generally assigned a Middle Pleistocene age, probably equivalent to the Late Biharian of Europe at 600,000 to 400,000 years ago (Butzer and Isaac 1975:892).

Lantian Man of Shensi province (Woo 1964, 1966) has also been attributed to *Homo erectus* (*Sinanthropus* species). Two specimens are represented: a man-

228 *Homo erectus*

dible recovered from Chenchiawo locality in 1963 and a skull from Kungwangling locality recovered in 1964. The skull has been reported to be more primitive in certain features than the Peking and Java *H. erectus* fossils, while the mandible has been aligned with those from Chou Kou Tien Locality 1 (Woo 1964, 1966; Aigner and Laughlin 1973). There is some confusion regarding the dating of the two fossils. On the basis of faunal evidence, Aigner and Laughlin (1973) suggest that the skull may be anywhere from 100,000 to 400,000 years older than the mandible. Chinese scholars, on the other hand, view the two specimens as 'roughly contemporaneous' (Chang 1977). In both instances, however (see also Aigner 1978), the skull is relegated to the 'early phase' of the Middle Pleistocene and the mandible is placed in the 'middle phase.' A New China Agency news release, dated November 28, 1976, reported Lantian Man as having been assigned an age of 600,000 years on the basis of paleomagnetism.[1] Presumably, the reference is to the skull since the Kungwangling stratum has tentatively been correlated with deposits slightly postdating the Brunhes-Matuyama magnetic reversal of 0.69 million years (Chang 1977).

Other purported *Homo erectus* fossils in China have been announced for Hupeh province and for Yunnan province (Chang 1977). The Hupeh materials (Ch'angyang) include a maxilla and tooth that were recovered in 1956 and possibly date to the 'late phase' of the Middle Pleistocene. The New China Agency news release, referred to above, reported the additional discovery in Hupeh in 1975 of three teeth attributed to the Yunshien apeman and noted that they were probably of the same geological age as Lantian Man. As mentioned by Howells (this volume), the Yuan-mou incisors (Yunnan province) are said to resemble teeth from Chou Kou Tien Locality 1 and have been assigned an age of 1.7 million years by Chinese researchers. At present, the evidence from both Hupeh and Yunnan may best be regarded as highly tentative for taxonomic considerations.

The Mapa calvarium (Kwangtung province), also mentioned by Howells, has variously been termed Neanderthal-like and Peking Man-like, with the original investigators (Woo and Peng 1959) noting a close similarity to *Homo soloensis*. In a recent morphological assessment of Neanderthal-like fossils, Santa Luca (1978) concludes that the only appropriate label for Mapa is *incertae sedis*. The specimen has been attributed to the late phase of the Middle Pleistocene (Chang 1977) and, faunally, to a late Riss/mid-Würm equivalent (Aigner 1978).

---

1 The news release was kindly forwarded to me by the Department of External Affairs, Government of Canada. It has since been reprinted in its entirety in the *Canadian Association for Physical Anthropology Newsletter* 2, no. 1 (February 1978).

JAVA

Jacob (this volume) sees three separate species of *Pithecanthropus* in Indonesia and prefers to view Peking Man and Lantian Man as two additional, separate species with that generic designation. Many other students, however, attribute the Middle Pleistocene Trinil zone fossils *P. erectus* to *Homo erectus* and Jacob's *Pithecanthropus* specimens in general to *Homo*. The *H. erectus* materials date from 0.9 to 0.6 million years.

There is increasing evidence that fossils associated with the Ngandong faunal zone, colloquially referred to as Solo Man (Jacob's *P. soloensis* and others' *Homo soloensis*), should be attributed to *homo erectus*. Recent cranial morphological studies suggest strong resemblances to the Trinil zone material (Santa Luca 1976, 1977, 1978), and a newly published study of endocranial casts (Holloway 1980) supports the view that the Solo crania are more closely related to earlier Indonesian *Homo erectus* than to Neanderthals. The connection seems further reinforced by Jacob's own morphological attribution to the Solo group of the Sangiran 17 skull from the Kabuh Beds of the Trinil zone. The most recent age for the Ngandong material is given by Jacob as 200,000 years, thus intimating a duration of possibly 700,000 years for *Homo erectus* in Indonesia.

Fossils ascribed to the Lower Pleistocene of Sangiran and Perning, assigned to the Djetis zone, have been described as *P. modjokertensis* by Jacob and *H. erectus*-like by others. Jacob accepts a date of 1.9 ± 0.4 million years, although this date has been questioned (Butzer and Isaac 1975). In view of the early date it is of interest that one of the Djetis-zone fossils, *Pithecanthropus* IV, is placed with australopithecine rather than with *Homo erectus* fossils on the basis of cranial shape analysis (Corruccini 1974). Others also view the *H. modjokertensis* fossils in pre-*erectus* terms, in this case bearing strong resemblances to *Homo habilis* of Olduvai Gorge (Tobias and von Koenigswald 1964; see also Boaz and Howell 1977), who has otherwise been regarded as an advanced *Australopithecus* (=*Homo*) *africanus*.

EAST AFRICA

East African fossils attributed to *Homo erectus* include KNM-ER 3733 and KNM-ER 3883 from the East Turkana basin, and OH 9 and OH 12 from Olduvai Gorge (Walker; Rightmire, this volume). By virtue of their occurrence in the Upper Member of the Koobi Fora Formation, the East Turkana skulls date somewhere between 1.9 and 1.5 million years. The basal date now appears to be fixed by newly published geochronological estimates for the KBS tuff. Gleadow (1980) has provided a new age of 1.87 ± 0.04 million years on the basis of

230 *Homo erectus*

refined techniques of fission track dating, while McDougall et al. (1980) have simultaneously arrived at a potassium-argon date of 1.89 ± 0.01 million years. According to these authors, the older dates of 2.4 million and 2.6 million years appear to have derived from contaminated samples.

Olduvai hominids OH 9 and OH 12 have been dated at 1.2 million and at 0.8 to 0.6 million years respectively on the basis of paleomagnetism and relative strata thickness estimates of the Olduvai Gorge geological beds (Rightmire, this volume and 1979). Day (1977:152-4) lists other finds from Olduvai Gorge that have been or might be attributed to *Homo erectus*: OH 2 (vault fragments), OH 15 (three teeth), OH 22 (partial mandible), OH 23 (mandibular fragment), OH 28 (left innominate and left femur shaft; assigned to *Homo erectus* by Day (1971)), OH 29 (three teeth and a phalange), OH 36 (ulna), and OH 51 (partial mandible with two teeth). Several of these specimens (OH 15, OH 22, OH 23, OH 29 with one of the teeth reassigned to OH 59, and OH 51) and one other specimen, OH 11 (a left maxilla fragment), have recently been described in detail by Rightmire (1980). Of the three mandibular fragments, he notes many similarities between the OH 22 specimen and *Homo erectus* jaws from Chou Kou Tien Locality 1 and from Ternifine in North Africa. Rightmire suggests a minimum age of 0.6 million years for the fossils of Bed IV which contained OH 22.

One other fossil from East Africa that has been attributed to *Homo erectus* is the cranium from Lake Ndutu in the Serengeti Plains (Clarke 1976). Its advanced features, however, have prompted others to regard Ndutu as a possible primitive *Homo sapiens* (Stringer, Howell, and Melentis 1979). The deposits in which the skull was found appear to be equivalent to the Masek Beds of Olduvai Gorge (Mturi 1976), which have been assigned an age of 0.6 to 0.4 million years (Hay 1976). A date of 0.6 to 0.5 million years has also been suggested on the basis of racemization of bone associated with the cranium (Mturi 1976).

SOUTH AFRICA

Two specimens from South Africa, specifically from the Swartkrans site, have been alluded to in this volume as possibly *Homo erectus*. They are the *Telanthropus* mandible and the partial SK-847 cranium. Illustrating the latter, Walker has noted similarities in this specimen to the KNM-ER 3733 cranium from East Turkana.

In a recent review of the hominids from Swartkrans, Kromdraai, Sterkfontein, and Makapansgat, Olson (1978) places all specimens in either *Homo* or *Paranthropus*, using the generic *Homo* to also include all australopithecines that are not the generally recognized 'robust' form (i.e. *Paranthropus*). Of the specimens assigned to *Homo*, four fossils, all from Swartkrans and all regarded as representing a single individual, are attributed to *Homo erectus* following the lead of Robinson (1961). These fossils include the *Telanthropus* jaw (SK-15) but not the SK-847 cranium. The latter is instead referred to *Homo (Australopithecus) africanus*, thus indicating a subgeneric as well as specific distinction from *Homo erectus*. Olson, however, does identify features that SK-847 has in common with accepted *Homo erectus* forms. The combination of primitive and advanced features are also seen in KNM-ER 1470 from East Turkana and in OH 24 (*Homo habilis*) from Olduvai Gorge, both of which Olson also refers to *Homo africanus* (a taxonomic designation, at least in terms of species, that is also ascribed to by Walker in this volume for the ER 1470 skull). In doing so, Olson recognizes that others may consider all three specimens as an intermediate morphotype (called either *H. habilis* or *H. modjokertensis*) between *H. africanus* and *H. erectus*.

Citing the work of others, Olson places the South African *Homo erectus* at 500,000 years or less and the SK-847 cranium at 2.0 to 1.5 million years.

NORTH AFRICA

There appears to be a consensus that the fossil remains of Ternifine, Algeria, mainly consisting of mandibles and formerly called *Atlanthropus mauritanicus*, represent *Homo erectus*. These materials have been assigned an age of around 700,000 to 500,000 years ago, while *A. mauritanicus* fossils from Morocco, the Thomas 1 and 2 specimens, are placed somewhat later at greater than 200,000 to, perhaps, 400,000 years ago (Jaeger, this volume).

Jaeger has grouped the still later specimens from Morocco, Sidi Abderrhamen, Rabat, and Salé, as *Homo erectus* but has also noted that they show several 'unexpected' modern features. These features and, perhaps, the late dating of the specimens, ca 220,000 to 150,000 years BP, are apparently what have led others to regard at least the Rabat and Salé fossils in frames of reference other than *Homo erectus*. In this volume, for example, Howells sees Salé as a late 'primitive,' and Thoma sees affinities between Rabat and certain European fossils such as Arago, Steinheim, Swanscombe, and Biache. Salé and Rabat may, therefore, fit well into the 'anteneanderthal' grouping.

EUROPE

The situation concerning the presence or absence of *Homo erectus* in Europe is far from clear. De Lumley and Howell, who together in this volume have re-

232 Homo erectus

viewed the European Middle Pleistocene fossils, maintain that they are distinct from East Asian and African *Homo erectus* forms, and note features reminiscent of Neanderthals or modern *Homo sapiens*.

The various fossils are not morphologically uniform. Some, such as Steinheim, Swanscombe, and Fontéchevade, show clearly progressive features that direct them toward Neanderthals (Howells, this volume; see also Santa Luca (1978) who characterizes Steinheim and Swanscombe as likely candidates for Neanderthal ancestry), while others, such as Vértesszöllös, Bilzingsleben, Petralona, and Mauer, show features that might plausibly align them with *Homo erectus*.

For the Vértesszöllös occipital, possibly dating to 500,000 years or more, the *H. erectus* connection has been illustrated by Thoma in this volume, although he would phylogenetically assign the specimen to *Homo sapiens*, emphasizing its intermediate morphology. Mania and Vlček (this volume) group Vértesszöllös with their find at Bilzingsleben (as does Thoma), which they regard as a subspecific variant of *Homo erectus*. The racemization date of 230,000 BP cited by those authors would now appear to be confirmed by a newly reported thorium-uranium date of $228,000^{+17,000}_{-20,000}$ BP on the Bilzingsleben travertine (Harmon, Głazek, and Nowak 1980).

The Petralona skull is enigmatic. A multivariate analysis of cranial shape (Corruccini 1974) has strongly supported Hemmer's (1972) assignment of the specimen to *Homo erectus*. In a new study, however, Stringer, Howell, and Melentis (1979), emphasizing craniofacial morphology, refer the fossil to a primitive grade of *Homo sapiens*. They further propose a *Homo sapiens* grade structure to accommodate virtually all European Middle Pleistocene fossils, as well as some others. Petralona is placed in the first of four grades together with Vértesszöllös, Bilzingsleben, and Mauer, as well as the Broken Hill specimen of sub-Saharan Africa. The North African Rabat and Salé specimens are less certainly referred to this grade as are, interestingly enough, the Ngandong Solo crania, otherwise attributed to *Homo erectus* (see above). Clearly, Petralona as well as Vértesszöllös and Bilzingsleben may yet be regarded as serious contenders for *Homo erectus* status. Petralona is now assigned an age of ca 300,000 years (Stringer, Howell, and Melentis 1979).

OTHER FOSSILS

Other fossils of note that have been reported on or mentioned in this volume are Eyasi I, Broken Hill, Saldanha (otherwise known as Hopefield or Elandsfontein), Omo II, and Garusi—all African finds. Most researchers now appear to agree that the first four represent archaic *Homo sapiens rhodesiensis* (Rightmire 1976; Protsch, this volume). These specimens range in age from possibly 130,000 BP for Omo II to 34,000 BP for Eyasi I (Howells, Protsch, this volume). Other dating techniques, however, detailed in Oakley, Campbell, and Molleson (1977), suggest that the four fossils might also be contemporaneous at ca 40,000 BP and, therefore, more than very late with respect to accepted *Homo erectus* forms.

Protsch (this volume) has associated the Garusi hominid, morphologically and mineralogically, with the hominids of Laetolil, south of Olduvai Gorge, which have been radiometrically dated at 3.8-3.6 million years (Leakey et al. 1976). A new taxon, *Australopithecus afarensis*, has recently been proposed to accommodate the Laetolil hominids together with fossil hominids recovered from the Hadar in the Afar triangle of Ethiopia, dated at 3.3-2.6 million years (Johanson, White, and Coppens 1978; Johanson and White 1979). The new species is regarded as more primitive than either *Australopithecus* (=*Homo*) *africanus* and *Australopithecus* (=*Paranthropus*) *robustus* or *Homo habilis* (=?*modjokertensis*) and *Homo erectus*, but demonstrating distinctive characteristics that suggest an ancestral relation to those groups.

SUMMARY

I have attempted to present this synopsis as objectively as possible, essentially counting the 'votes' of the various investigators, in this volume and elsewhere, who have worked directly with the fossil evidence. The results of the 'election' might be summarized as in the chart, where fossils within each of the Old World areas discussed are morphologically grouped against a background of absolute age. The solid brackets with the labels of *H. erectus* and *H. sapiens* indicate a reasonable certainty (consensus?) with respect to the separate groupings, while the broken brackets, with or without labels, indicate uncertainty. The cross-bars at the ends of the brackets denote currently known temporal limits. They have been left off the brackets encompassing the very early and late Chinese fossils and off the upper limits of the brackets encompassing SK-15 of South Africa.

There is uncertainty as to whether Ndutu in East Africa and Salé-Sidi Abderrhamen-Rabat in North Africa, because of their apparently advanced features, should be included in *Homo erectus*. The fact that some investigators do regard them as such is denoted by the connection of their brackets with those of the more certain preceding fossils. In Europe, there is uncertainty surrounding the taxonomic status of Bilzingsleben, Petralona, Vértesszöllös, and Mauer.

Pre-*H. erectus* forms have been labelled and included in the chart for Java, East Africa, and South Africa, as discussed in this review. Their morphological-cum-taxonomic status is in need of further study. In that connection, the Yuanmou discovery in China, as well as possible australopithecine discoveries there (see Chang 1977), offers intriguing possibilities for exploring the origins and widespread occurrence of 'true' *Homo* throughout the Asian-African theater during the Lower Pleistocene.

234 *Homo erectus*

ABSOLUTE AGE (YRS BP)	CHINA	JAVA
100,000	MAPA CH'ANGYANG	
200,000		⎡
300,000		SOLO
400,000	⎡	
500,000	PEKING (H. erectus)	(H. erectus)
600,000	LANTIAN ⎦	
700,000		TRINIL ZONE
800,000		
900,000		⎦
1,000,000		
1,100,000		
1,200,000		
1,300,000		
1,400,000		
1,500,000		
1,600,000		⎡
1,700,000	YUAN MOU	
1,800,000		PITHECANTHROPUS IV
1,900,000		MODJOKERTO
2,000,000		(Pre-erectus) ⎦

A framework of time, space, and morphology for select Pleistocene fossils, focusing on *Homo erectus*. (Thanks to David W. Laverie of the Archaeological Survey of Canada, National Museum of Man, for his advice and drafting skills in the production of this chart.)

235 J.S. Cybulski

E. AFRICA	S. AFRICA	N. AFRICA	EUROPE
H. sapiens [EYASI I, OMO II]	*H. sapiens* [SALDANHA, BROKEN HILL]		*H. sapiens* [FONTÉCHEVADE, STEINHEIM, SWANSCOMBE]
		[SALÉ, SIDI ABDÉRRHAMEN, RABAT	
		THOMAS 1&2]	*H. erectus*? [BILZINGSLEBEN, PETRALONA
[NDUTU, OH 12, OH 22, OH 28	*H. erectus* [SK-15]	*H. erectus* [TERNIFINE]	VÉRTESSZŐLLŐS, MAUER]
H. erectus OH 9			
[ER 3733, ER 3883]	[SK-847] Pre-erectus		
Pre-erectus [OH 24, ER 1470]			

The distribution and current status of fossil hominids in the Middle Pleistocene certainly suggests that *Homo erectus* was a widespread species in the Old World, and even Europe cannot definitely be eliminated from that distribution of the basis of present interpretations. The time limits currently set in each area are by no means inviolable and will undoubtedly become more precise, and perhaps altered, with further research. Above all, we must not underestimate the range of morphological variability that may have been inherent in a species such as *Homo erectus* as we attempt to refine and further define its status in the evolution of man.

MAP SHOWING FOSSIL SITES

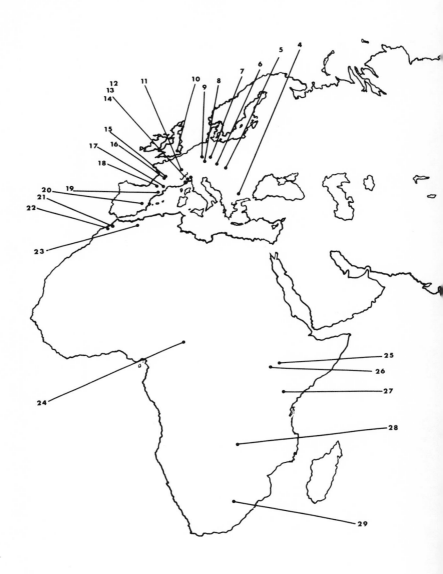

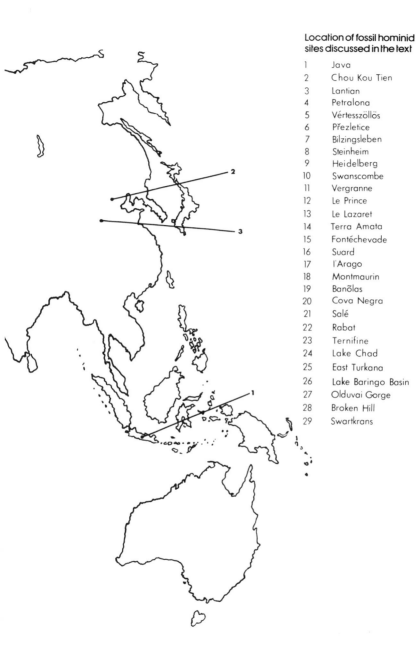

Location of fossil hominid sites discussed in the text

1. Java
2. Chou Kou Tien
3. Lantian
4. Petralona
5. Vértesszöllös
6. Přezletice
7. Bilzingsleben
8. Steinheim
9. Heidelberg
10. Swanscombe
11. Vergranne
12. Le Prince
13. Le Lazaret
14. Terra Amata
15. Fontéchevade
16. Suard
17. l'Arago
18. Montmaurin
19. Banõlas
20. Cova Negra
21. Salé
22. Rabat
23. Ternifine
24. Lake Chad
25. East Turkana
26. Lake Baringo Basin
27. Olduvai Gorge
28. Broken Hill
29. Swartkrans

Bibliography

Aguirre, E., J.M. Basabe, and T. Torres. 1976. Los fosiles humanos de Atapuerca (Burgos): nota preliminar. *Zephyrus* 26-7:489-511

Aigner, J.S. 1978. The paleolithic of China. In *Early man in America from a circum-Pacific perspective*, edited by A.L. Bryan, pp. 25-41. Edmonton: Archaeological Researches International

Aigner, J.S. and W.S. Laughlin. 1973. The dating of Lantian Man and his significance for analyzing trends in human evolution. *American Journal of Physical Anthropology* 39:97-110

Alimen, H. 1975. Les 'isthmes' hispano-marocain et siculo-tunisien aux temps acheuléens. *L'Anthropologie* 79:399-436

Andersson, J.G. 1934. *Children of the yellow earth: studies in prehistoric China*. London: Kegan Paul, Trench, Trubner & Co. Ltd

Andree, J. 1939. *Der eiszeitliche Mensch in Deutschland und seine Kulturen*. Stuttgart: F. Enke

Andrews, R.C. 1945. *Meet your ancestors: a biography of primitive man*. New York: Viking

Arambourg, C. 1938. Mammifères fossiles du Maroc. *Mémoires de la Société des sciences naturelles et physiques du Maroc* 46:1-74

— 1956. Le gisement pléistocène de Ternifine et l'*Atlanthropus*. *Bulletin de la Société belge de géologie, paléontologie et hydrologie* 65:132-6

— 1957. Récentes découvertes de paléontologie humaine réalisées en Afrique du Nord française (l'*Atlanthropus* de Ternifine — l'hominien de Casablanca). In *Proceedings of the Third Pan-African Congress on Prehistory*, Livingstone, 1955, edited by J.D. Clark, pp. 186-94. London: Chatto and Windus

— 1960. Au sujet de *Elephas iolensis* Pomel. *Bulletin d'archéologie marocaine* 3:93-105

- 1963. Le gisement de Ternifine: l'*Atlanthropus* de Ternifine. *Archives de l'Institut de paléontologie humaine* 32:37-109. Paris
- 1969-70. Les vertébrés du pléistocène de l'Afrique du Nord. *Archives du Muséum national d'histoire naturelle* 10:1-126. Paris

Arambourg, C., J. Arènes, and G. Depape. 1953. Contribution à l'étude des flores fossiles quaternaires de l'Afrique du Nord. *Archives du Muséum national d'histoire naturelle*, Série 7, 2:1-85. Paris

Arambourg, C., and P. Biberson. 1955. Découverte de vestiges humains acheuléens dans la carrière Sidi Abd-er-Rahman prés Casablanca. *Comptes-rendus de l'Academie des sciences, Paris* 240:1661-3
- 1956. The fossil human remains from the Paleolithic site of Sidi Abderrahman (Morocco). *American Journal of Physical Anthropology* 14:467-89

Arambourg, C., and R. Hoffstetter. 1954. Découverte en Afrique du Nord, de restes humains du paléolithique inférieur. *Comptes-rendus de l'Academie des sciences, Paris* 239:72-4
- 1955. Le gisement de Ternifine: résultats des fouilles de 1953 et découverte de nouveaux restes d'*Atlanthropus*. *Comptes-rendus de l'Academie des sciences, Paris* 241:431-3
- 1963. Le gisement de Ternifine: historique et géologie. *Archives de l'Institut de paléontologie humaine* 32:1-36. Paris

Ariëns Kappers, C.U. 1929. The fissures on the frontal lobes of *Pithecanthropus erectus* Dubois compared with those of Neanderthal men, *Homo recens* and Chimpanzee. *Proceedings Koninklijke Akademie van Wetenschappen, Amsterdam* 32:182-95
- 1936. The endocranial casts of the Ehringsdorf and *Homo soloensis* skulls. *Journal of Anatomy* 71:61-76

Bada, J.L., R. Protsch, and R. Berger. 1974. Concordance of collagen-based radiocarbon and aspartic-acid racemization ages. *Proceedings of the National Academy of Sciences* 71:914-17

Barbour, G.B. 1935. Memorial of Davidson Black. *Proceedings of the Geological Society of America* for 1934, pp. 193-202

Beaudet, G. 1969. Le plateau central marocain et ses bordures; étude géomorphologique. Thèse, Université de Paris. Rabat: Imprimeries françaises et marocaines

Beaudet, G., G. Maurer, and A. Ruellan. 1967. Le quaternaire marocain; observations et hypothèses nouvelles. *Revue de géographie physique et de géologie dynamique*, Série 2, 9:269-309. Paris

Behrensmeyer, A.K. 1970. Preliminary geological interpretation of a new hominid site in the Lake Rudolf basin. *Nature* 226:225-6

- 1975. The taphonomy and paleoecology of Plio-Pleistocene vertebrate assemblages east of Lake Rudolf, Kenya. *Bulletin of the Museum of Comparative Zoology, Harvard*, 146:473-578
- 1976. Fossil assemblages in relation to sedimentary environments in the East Rudolf succession. In *Earliest man and environments in the Lake Rudolf basin: stratigraphy, paleoecology, and evolution*, edited by Y. Coppens, F.C. Howell, G.Ll. Isaac, and R.E.F. Leakey, pp. 383-401. Chicago and London: University of Chicago Press

Beveniste, R.E. and G.J. Rodaro. 1976. Evolution of type C viral genes: evidence for an Asian origin of man. *Nature* 261:101-8

Biberson, P. 1956. Le gisement de l'Atlanthrope de Sidi Abderrahman (Casablanca). *Bulletin d'archéologie marocaine* 1:39-92
- 1961. Le cadre paléogéographique de la préhistoire du Maroc atlantique. *Publications de la Service des antiquités du Maroc* 16:1-235. Rabat
- 1971. Essai de redéfinition des cycles climatiques du quaternaire continental du Maroc. *Bulletin de l'Association française pour l'étude du quaternaire* 1(26):3-13. Paris

Birkner, F. 1937. Reste des urmenschen in Afrika? *Germania*

Black, D. 1927. On a lower molar hominid tooth from the Chou Kou Tien deposit. *Palaeontologia Sinica*, Series D, 7(1)
- 1929a. Preliminary note on additional *Sinanthropus* material discovered in Chou Kou Tien during 1928. *Bulletin of the Geological Society of China* 8:15-32
- 1929b. Preliminary notice of the discovery of an adult *Sinanthropus* skull at Chou Kou Tien. *Bulletin of the Geological Society of China* 8:207-30
- 1930a. Interim report on the skull of *Sinanthropus*. *Bulletin of the Geological Society of China* 9:7-22
- 1930b. Notice of the recovery of a second adult *Sinanthropus* skull specimen. *Bulletin of the Geological Society of China* 9:97-100
- 1931. On an adolescent skull of *Sinanthropus pekinensis* in comparison with an adult skull of the same species and with other hominid skulls, recent and fossil. *Palaeontologia Sinica*, Series D, 7(2)

Black, D., P. Teilhard de Chardin, C.C. Young, and W.C. Pei. 1933. Fossil man in China; the Choukoutien cave deposits with a synopsis of our present knowledge of the Late Cenozoic in China. *Memoirs of the Geological Survey of China*, Series A, 11

Boaz, N.T. and A.K. Behrensmeyer. 1976. Hominid taphonomy: transport of human skeletal parts in an artificial fluviatile environment. *American Journal of Physical Anthropology* 45:53-60

Boaz, N.T. and F.C. Howell. 1977. A gracile hominid cranium from Upper Member G of the Shungura Formation, Ethiopia. *American Journal of Physical Anthropology* 46:93-108

Bowen, B.E. and C.F. Vondra. 1973. Stratigraphical relationships of the Plio-Pleistocene deposits, East Rudolf, Kenya. *Nature* 242:391-3

Bowers, J.Z. 1972. *Western medicine in a Chinese palace: Peking Union Medical College, 1917-1951.* New York: Josiah Macy Jr Foundation

Brace, C.L. 1967. *The stages of human evolution: human and cultural origins.* Englewood Cliffs: Prentice Hall

— 1973. Sexual dimorphism in human evolution. *Yearbook of Physical Anthropology* 16:31-49

Brain, C.K. 1970. New finds at the Swartkrans australopithecine site. *Nature* 225:1112-19

Brébion, Ph. 1973. Nouvelles recherches sur les Gastéropodes pliocènes et quaternaires du Maroc atlantique. *Comptes-rendus de l'Academie des sciences, Paris,* Série D, Sciences Naturelles 277:489-92

— 1976. Etude biostratigraphique des Gastéropodes du quaternaire marin du Maroc atlantique. *Comptes-rendus de l'Academie des sciences, Paris,* Série D, Sciences Naturelles 283:1579-82

Briggs, L.C. 1968. Hominid evolution in northwest Africa and the question of the North African 'neanderthaloids.' *American Journal of Physical Anthropology* 29:377-86

Brock, A. and G.Ll. Isaac. 1974. Paleomagnetic stratigraphy and chronology of hominid-bearing sediments east of Lake Rudolf, Kenya. *Nature* 247:344-8

— 1976. Reversal stratigraphy and its application at East Rudolf. In *Earliest man and environments in the Lake Rudolf basin: stratigraphy, paleoecology, and evolution,* edited by Y. Coppens, F.C. Howell, G.Ll. Isaac, and R.E.F. Leakey, pp. 148-62. Chicago and London: University of Chicago Press

Broom, R and J.T. Robinson. 1950. Sterkfontein ape-man, *Plesianthropus. Transvaal Museum Memoir* 4(1)

— 1952. Swartkrans ape-man; *Paranthropus crassidens. Transvaal Museum Memoir* 6

Broom, R. and G.W.H. Schepers. 1946. The South African fossil ape-men; the Australopithecinae. *Transvaal Museum Memoir* 2

Brown, F.H. and R. T. Shuey. 1976. Magnetostratigraphy of the Shungura and Usno formations, Lower Omo Valley, Ethiopia. In *Earliest man and environments in the Lake Rudolf basin: stratigraphy, paleoecology, and evolution,* edited by Y. Coppens, F.C. Howell, G. Ll. Isaac, and R.E.F. Leakey, pp. 64-78. Chicago and London: University of Chicago Press

Burr, D.B. 1976. Neandertal vocal tract reconstructions: a critical appraisal. *Journal of Human Evolution* 5:285-90

Butzer, K.W. 1975. Pleistocene littoral-sedimentary cycles of the Mediterranean basin: a Mallorquin view. In *After the australopithecines: stratigraphy, ecology and culture change in the Middle Pleistocene*, edited by K.W. Butzer and G.Ll. Isaac, pp. 25-71. The Hague and Paris: Mouton

Butzer, K.W. and G. Ll. Isaac, editors. 1975. *After the australopithecines: stratigraphy, ecology and culture change in the Middle Pleistocene.* The Hague and Paris: Mouton

Campbell, B.G. 1963. Quantitative taxonomy and human evolution. In *Classification and human evolution*, edited by S.L. Washburn, pp. 50-74. Chicago: Aldine

— 1965. The nomenclature of the Hominidae including a definitive list of hominid taxa. *Royal Anthropological Institute Occasional Paper* 22

— 1972. Conceptual progress in physical anthropology: fossil man. *Annual Review of Anthropology* 1:27-54

Campy, M., J. Chaline, C. Guérin, and B. Vandermeersch. 1974. Une canine humaine associée à une faune Mindel récent dans le remplissage de l'Aven de Vergranne (Doubs). *Comptes-rendus de l'Academie des sciences, Paris*, Série D, Sciences Naturelles 278:3187-90

Carney, J., A. Hill, J.A. Miller, and A. Walker. 1971. Late australopithecine from Baringo District, Kenya. *Nature* 230-509-14

Cerling, T.E. 1976. Oxygen-isotope studies of the East Rudolf volcanoclastics. In *Earliest man and environments in the Lake Rudolf basin: stratigraphy, paleoecology, and evolution*, edited by Y. Coppens, F.C. Howell, G.Ll. Isaac, and R.E.F. Leakey, pp. 105-14. Chicago and London: University of Chicago Press

Cerling, T.E., D.L. Biggs, C.F. Vondra, and H. Sveck. 1975. Use of oxygen isotope ratios in correlation of tuffs, East Rudolf basin, northern Kenya. *Earth and Planetary Science Letters* 25:291-6

Chang, K.C. 1977. Chinese palaeoanthropology. *Annual Review of Anthropology* 6:137-59

Chia, L.P. 1975. *The cave home of Peking Man.* Peking: Foreign Languages Press

Chiu, C.L., Y.M. Gu, Y.Y. Zhang, and S.S. Chang. 1973. Newly discovered *Sinanthropus* remains and stone artefacts at Choukoutien. *Vertebrata Palasiatica* 11(2):109-31

Choubert, G. and A. Faure-Muret. 1971. Nouvelle contribution à l'étude radiométrique du quaternaire du Maroc. *Quaternaria* 14:205-7

Choubert, G., A. Faure-Muret, and G.C. Maarleveld. 1967. Nouvelles dates isotopiques du quaternaire marocain et leur signification. *Comptes-rendus de l'Academie des sciences, Paris*, Série D, Sciences Naturelles 264:434-7

Clarke, R.J. 1976. New cranium of *Homo erectus* from Lake Ndutu, Tanzania. *Nature* 262:485-7

Combier, J. 1971. Le gisement prémoustérien et acheuléen d'Orgnac. *Etudes préhistoriques* 1(mars):24-6

Coon, C.S. 1962. *The origin of races*. New York: Alfred A. Knopf

Coque, R. 1962. *La tunisie présaharienne; étude géomorphologique*. Paris: Armand Colin

Coque, R. and A. Jauzein. 1965. Le quaternaire moyen de l'Afrique du Nord. *Bulletin de l'Association française pour l'étude du quaternaire* 2:117-32. Paris

Corruccini, R.S. 1974. Calvarial shape relationships between fossil hominids. *Yearbook of Physical Anthropology* 18:89-109

— 1975. Metrical analysis of Fontéchevade II. *American Journal of Physical Anthropology* 42:95-7

Curtis, G.H., R. Drake, T.F. Cerling, and J.H. Hampel. 1975. Age of KBS tuff in Koobi Fora formation, East Rudolf, Kenya. *Nature* 258:395-8

Darwin, C. 1874. *The descent of man and selection in relation to sex*. New York: A.L. Burt. Reprinted from the 2nd English ed., rev. and aug.

Day, M.H. 1971. Postcranial remains of *Homo erectus* from Bed IV, Olduvai Gorge, Tanzania. *Nature* 232:383-7

— 1972. The Omo human skeletal remains. In *The origin of Homo sapiens*, edited by F. Bordes, pp. 31-5. Paris: UNESCO

— 1977. *Guide to fossil man; a handbook of human palaeontology*. Third edition. Chicago and London: University of Chicago Press

Day, M.H. and R.E.F. Leakey. 1973. New evidence of the genus *Homo* from East Rudolf, Kenya (I). *American Journal of Physical Anthropology* 39:341-54

— 1974. New evidence of the genus *Homo* from East Rudolf, Kenya (III). *American Journal of Physical Anthropology* 41:367-80

Day, M.H., R.E.F. Leakey, A.C. Walker, and B.A. Wood. 1975. New hominids from East Rudolf, Kenya (I). *American Journal of Physical Anthropology* 42:461-76

— 1976. New hominids from East Turkana, Kenya. *American Journal of Physical Anthropology* 45:369-436

Day, M.H. and T.I. Molleson. 1973. The Trinil femora. In *Human evolution*, edited by M.H. Day, pp. 127-54. London: Taylor and Francis

DeVore, I. and K.R.L. Hall. 1965. Baboon ecology. In *Primate behavior: field studies of monkeys and apes*, edited by I. DeVore, pp. 20-52. New York: Holt, Rinehart and Winston

Dobzhansky, T. 1944. On species and races of living and fossil man. *American Journal of Physical Anthropology* 2:251-65

Dubois, E. 1894. *Pithecanthropus erectus, eine menschenaehnliche uebergangsform aus Java*. Batavia: Landesdruckerei

— 1932. The distinct organization of *Pithecanthropus* of which the femur bears evidence, now confirmed from other individuals of the described species. *Proceedings Koninklijke Akademie van Wetenschappen, Amsterdam* 35:716-22

— 1936. Racial identity of *Homo soloensis* Oppenoorth (including *Homo modjokertensis* von Koenigswald) and *Sinanthropus pekinensis* Davidson Black. *Proceedings Koninklijke Akademie van Wetenschappen, Amsterdam* 39:1180-5

Duplessy, J.-C., C. Vergnaud-Grazzini, G. Delibrias, C. Lalou, and R. Letolle. 1976. Paléoclimatologie des temps quaternaires à l'aide des méthodes nucléaires. In *La préhistoire française I: les civilisations paléolithiques et mésolithiques de la France*, edited by H. de Lumley, pp. 352-61. Paris: Centre national de la recherche scientifique

Eldredge, N. and S.J. Gould. 1972. Punctuated equilibria: an alternative to phyletic gradualism. In *Models in paleobiology*, edited by T.J.M. Schopf, pp. 82-115. San Francisco: Freeman, Cooper

Ennouchi, E. 1962a. Un crâne d'homme ancien au Jebel Irhoud. *Comptes-rendus de l'Academie des sciences, Paris* 254:4330-2

— 1962b. Un néandertalien: l'homme du Jebel Irhoud (Maroc). *L'Anthropologie* 66:279-98

— 1966. Essai de datation du gisement du Jebel Irhoud (Maroc). *Comptes-rendus sommaire de la Société géologique de France* 10:405-6 Paris

— 1968. Le deuxième crâne de l'homme d'Irhoud. *Annales de paléontologie (vertébrés)* 54:117-28. Paris

— 1969a. Présence d'un enfant néanderthalien au Jebel Irhoud (Maroc). *Annales de paléontologie (vertébrés)* 55:251-65. Paris

— 1969b. Les empreintes des cerveaux des néanderthaliens marocains. *Service de géologie, notes et mémoires* No. 213 (29):63-70. Rabat

— 1969c. Découverte d'un pithécanthropien au Maroc. *Comptes-rendus de l'Academie des sciences, Paris*, Série D, Sciences Naturelles 269:763-5

— 1970. Un nouvel archanthropien au Maroc. *Annales de paléontologie (vertébrés)* 56:95-107. Paris

- 1972. Nouvelle découverte d'un archanthropien au Maroc. *Comptes-rendus de l'Academie des sciences, Paris*, Série D, Sciences Naturelles 274:3088-90
- 1976. Un deuxième archanthropien à la carrière Thomas III (Maroc); étude préliminaire (5 juin 1972). *Bulletin du Muséum national d'histoire naturelle*, Série 3, Sciences de la Terre 56:273-96. Paris

Ferembach, D. 1976. Les restes humains artériens de Témara (campagne 1975). *Bulletins et mémoires de la Société d'anthropologie de Paris*, Série 13, 3:175-80

Findlater, I.C. 1976. Tuffs and the recognition of isochronous mapping units in the East Rudolf succession. In *Earliest man and environments in the Lake Rudolf basin: stratigraphy, paleoecology, and evolution*, edited by Y. Coppens, F.C. Howell, G.Ll. Isaac, and R.E.F. Leakey, pp. 94-104. Chicago and London: University of Chicago Press
- 1978a. Stratigraphy. In *Koobi Fora research project. Vol. 1: The fossil hominids and an introduction to their context 1968-1974*, edited by M. Leakey and R.E.F. Leakey. London: Oxford University Press
- 1978b. Isochronous surfaces within the Plio-Pleistocene sediments east of Lake Turkana. In *Geological background to fossil man*, edited by W.W. Bishop, pp. 415-20. Edinburgh: Scottish Academic Press

Fitch, F.J., I.C. Findlater, R.T. Watkins, and J.A. Miller. 1974. Dating of the rock succession contributing fossil hominids at East Rudolf, Kenya. *Nature* 251:213-15

Fitch, F.J., P.J. Hooker, and J.A. Miller. 1976. ^{40}Ar / ^{39}Ar dating of the KBS tuff in Koobi Fora formation, East Rudolf, Kenya. *Nature* 263:740-4
- 1978. Geochronological problems and radioisotopic dating in the Gregory Rift Valley. In *Geological background to fossil man*, edited by W.W. Bishop, pp. 441-61. Edinburgh: Scottish Academic Press

Fitch, F.J., and J.A. Miller. 1970. Radioisotopic age determinations of Lake Rudolf artefact site. *Nature* 226:226-8
- 1976. Conventional Potassium-Argon and Argon-40 / Argon-39 dating of volcanic rocks from East Rudolf. In *Earliest man and environments in the Lake Rudolf basin: stratigraphy, paleoecology, and evolution*, edited by Y. Coppens, F.C. Howell, G.Ll. Isaac, and R.E.F. Leakey, pp. 123-47. Chicago and London: University of Chicago Press

Flint, R.F. 1971. *Glacial and quaternary geology*. New York: John Wiley

Freeman, L.G. 1975. Acheulean sites and stratigraphy in Iberia and the Maghreb. In *After the australopithecines: stratigraphy, ecology and culture change in the Middle Pleistocene*, edited by K.W. Butzer and G.Ll. Isaac, pp. 661-743. The Hague and Paris: Mouton

Garn, S.M. and W.D. Block. 1970. The limited nutritional value of cannibalism. *American Anthropologist* 72:106

Gauss, G.F. 1934. *The struggle for existence*. Baltimore: Williams and Wilkins

Gigout, M. 1959. Ages, par radiocarbone, de deux formations des environs de Rabat (Maroc). *Comptes-rendus de l'Academie des sciences, Paris* 249:2802-3

Gleadow, A.J.W. 1980. Fission track age of the KBS tuff and associated hominid remains in northern Kenya. *Nature* 284:225-30

Gould, S.J. and N. Eldredge. 1977. Punctuated equilibria: the tempo and mode of evolution reconsidered. *Paleobiology* 3:115-51

Grimm, H., and D. Mania. 1976. Bilzingsleben B — ein weiterer mittelpleistozäner hominiden-fund aus dem Elbe–Saale–Gebiet. *Biologische Rundschau* 14:155-6

Grimm, H., V. Toepfer, and D. Mania. 1974. Ein neuer hominidenfund in Europa. Nachtrag zum Vorbericht über Bilzingsleben, Kr. Artern. *Zeitschrift für Archäologie* 8:175-6

Groves, C.P. and V. Mazak. 1975. An approach to the taxonomy of the Hominidae: gracile Villafranchian hominids of Africa. *Casopis pro Mineralogii a Geologii* 20(3):225-47

Harmon, R.S., J. Glazek, and K. Nowak. 1980. ^{230}Th / ^{234}U dating of travertine from the Bilzingsleben archaeological site. *Nature* 284:132-5

Hay, R.L. 1976. *Geology of the Olduvai Gorge: a study of sedimentation in a semiarid basin*. Berkeley: University of California Press

Hemmer, H. 1972. Notes sur la position phylétique de l'homme de Petralona. *L'Anthropologie* 76:155-61

Hennig, E. 1948. Quartarfaunen und ugeschicte Ostafrikas. *Naturwissenschaftliche Rundschau* 1:212-17

Hennig, W. 1966. *Phylogenetic systematics*. Urbana, Chicago, and London: University of Illinois Press

Holloway, R.L. 1975. Early hominid endocasts: volumes, morphology, significance for hominid evolution. In *Primate functional morphology and evolution*, edited by R.H. Tuttle, pp. 393-415

— 1980. Indonesian 'Solo' (Ngandong) endocranial reconstructions: some preliminary observations and comparisons with Neandertal and *Homo erectus* groups. *American Journal of Physical Anthropology* 53:285-95

Hood, D. 1964. *Davidson Black, a biography*. Toronto: University of Toronto Press

Howell, F.C. 1960. European and northwest African Middle Pleistocene hominids. *Current Anthropology* 1:195-232

Howells, W.W. 1959. *Mankind in the making: the story of human evolution*. New York: Doubleday

- 1966. Homo erectus. *Scientific American* 215(5):46-53
- 1977. Hominid fossils. In *Paleoanthropology in the People's Republic of China*, edited by W.W. Howells and P.J. Tsuchitani, pp. 66-78. Washington: National Academy of Sciences

von Huene, F. 1940. Die stammesgeschichtliche gestalt der Wirbeltiere — ein lebenslauf. *Paläontologische Zeitschrift* 22:55-62

Hurford, A.J. 1974. Fission-track dating of a vitric tuff from East Rudolf, Kenya. *Nature* 249:236-7

Hurford, A.J., A.J.W. Gleadow, and C.W. Naeser. 1976. Fission-track dating of pumice from the KBS tuff, East Rudolf, Kenya. *Nature* 263:738-40

Jacob, T. 1967. *Some problems pertaining to the racial history of the Indonesian region: a study of human skeletal and dental remains from several prehistoric sites in Indonesia and Malaysia.* Utrecht: Neerlandia
- 1972. The problem of head-hunting and brain-eating among Pleistocene men in Indonesia. *Archaeology and Physical Anthropology in Oceania* 7:81-91
- 1973. Palaeoanthropological discoveries in Indonesia with special reference to the finds of the last two decades. *Journal of Human Evolution* 2:473-85
- 1975a. Morphology and paleoecology of early man in Java. In *Paleoanthropology: morphology and paleoecology*, edited by R.H. Tuttle, pp. 311-25. The Hague: Mouton
- 1975b. L'homme de Java. *La Recherche* 6(62):1027-32
- 1975c. The pithecanthropines of Indonesia. *Bulletins et mémoires de la Société d'anthropologie de Paris*, Série 13, 2:243-56
- 1976a. Early populations in the Indonesian region. In *The origin of the Australians*, edited by R.L. Kirk and A.G. Thorne, pp. 81-93. Atlantic Highlands, N.J.: Humanities Press
- 1976b. A fossil endocranial cast from Sangiran: a preliminary report. Paper read at 9th International Congress of Prehistoric and Protohistoric Sciences, Nice

Jaeger, J.-J. 1969. Les rongeurs du pléistocène moyen de Ternifine (Algérie). *Comptes-rendus de l'Academie des sciences, Paris*, Série D, Sciences Naturelles 269:1492-5
- 1975a. Les Muridae (Mammalia, Rodentia) du pliocène et du pléistocène du Maghreb. Origine; evolution; données biogéographiques et paléoclimatiques. Thèse, Université de Montpellier. Paris: Centre national de la recherche scientifique
- 1975b. The mammalian faunas and hominid fossils of the Middle Pleistocene of the Maghreb. In *After the australopithecines: stratigraphy, ecology and culture change in the Middle Pleistocene*, edited by K.W. Butzer and G.Ll. Isaac, pp. 399-418. The Hague and Paris: Mouton

- 1975c. Découverte d'un crâne d'hominidé dans le pléistocène moyen du Maroc. *Centre national de la recherche scientifique, colloques internationaux, problèmes actuels de paléontologie (evolution des vertébrés)* 218:897-902. Paris
Janossy, D. 1975. Mid-Pleistocene microfaunas of continental Europe and adjoining areas. In *After the australopithecines: stratigraphy, ecology and culture change in the Middle Pleistocene*, edited by K.W. Butzer and G.Ll. Isaac, pp. 375-97. The Hague and Paris: Mouton
Janus, C.G. 1975. *The search for Peking Man*. New York: Macmillan
Johanson, D.C., and M. Taieb. 1976. Plio-Pleistocene hominid discoveries in Hadar, Ethiopia. *Nature* 260:293-7
Johanson, D.C. and T.D. White. 1979. A systematic assessment of early African hominids. *Science* 203:321-30
Johanson, D.C., T.D. White, and Y. Coppens. 1978. A new species of the genus *Australopithecus* (Primates: Hominidae) from the Pliocene of eastern Africa. *Kirtlandia* 28:1-14. Cleveland Museum of Natural History
Johnson, G.D. 1974. Cainozoic lacustrine stromatolites from the hominid-bearing sediments east of Lake Rudolf, Kenya. *Nature* 247:520-3
Johnson, G.D. and R.G.H. Raynolds. 1976. Late Cenozoic environments of the Koobi Fora formation: the Upper Member along the western Koobi Fora ridge. In *Earliest man and environments in the Lake Rudolf basin: stratigraphy, paleoecology, and evolution*, edited by Y. Coppens, F.C. Howell, G.Ll. Isaac, and R.E.F. Leakey, pp. 115-22. Chicago and London: University of Chicago Press
Kaufman, A., W.S. Broecker, T.L. Ku, and D.L. Thurber. 1971. The status of U-series methods of mollusk dating. *Geochimica and Cosmochimica Acta* 35:1155-83
Kay, R.F. and M. Cartmill. 1977. Cranial morphology and adaptations of *Palaechthon nacimienti* and other Paromomyidae (Plesiadapoidea, ? Primates), with a description of a new genus and species. *Journal of Human Evolution* 6:19-53
Keith, A. 1913. The Piltdown skull and brain cast. *Nature* 92:197-9, 292, 345-6
- 1920. *The antiquity of man*. London: Williams and Norgate
Kiszely, I. 1974. On the possibilities and methods of the chemical determination of sex from bones. *Ossa* 1:51-62
Kochetkova, V.I. 1970. Reconstruction de l'endocrâne de l'*Atlanthropus mauritanicus* et de l'*Homo habilis*. *Proceedings 8th International Congress of Anthropological and Ethnological Sciences* 1:102-4
von Koenigswald, G.H.R. 1931. Sinanthropus, Pithecanthropus, en de ouderdom van de Trinil lagen. *Overdrukt De Mijningenieur* 11:198-202

- 1935. Eine fossile Säugetierfauna mit simia aus Südchina. *Proceedings Koninklijke Akademie van Wetenschappen, Amsterdam* 38:872-9
- 1952. *Gigantopithecus blacki* von Koenigswald, a giant fossil hominoid from the Pleistocene of southern China. *Anthropological Papers of the American Museum of Natural History* 43:295-325
- 1955. *Begegnungen mit dem vormenschen*. Dusseldorf and Cologne: Eugen Diederichs
- 1975. Early man in Java: catalogue and problems. In *Paleoanthropology: morphology and paleoecology*, edited by R.H. Tuttle, pp. 303-9. The Hague: Mouton

von Koenigswald, G.H.R. and A.K. Ghosh. 1973. Stone implements from the Trinil beds of Sangiran, central Java. *Proceedings Koninklijke Nederlandse Akademie van Wetenschappen*, Series B, Physical Sciences 76:1-34

von Koenigswald, G.H.R. and F. Weidenreich. 1939. The relationship between *Pithecanthropus* and *Sinanthropus*. *Nature* 144:926-9

Kohl-Larsen, L. 1943. *Auf den spuren des vormenschen; forschungen, fahrten und erlebnisse in Deutsch-Ostafrika (Deutsche Afrika-expedition 1934-1936 und 1937-1939)*. Stuttgart: Strecker and Schröder

Kretzoi, M. and L. Vértes. 1965. Upper Biharian (Intermindel) pebble-industry occupation site in western Hungary. *Current Anthropology* 6:74-87

Kruuk, H. 1972. *The spotted hyena; a study of predation and social behavior*. Chicago: University of Chicago Press

Kuhn, T.S. 1970. *The structure of scientific revolutions*. Chicago and London: University of Chicago Press

Kurth, G. 1965. Die (eu)homininen. In *Menschliche abstammungslehre: fortschritte der "anthropogenie", 1863-1964*, edited by G. Heberer, pp. 357-425. Stuttgart: Gustav Fischer

Leakey, L.S.B. 1936. A new fossil skull from Eyassi, East Africa: discovery by a German expedition. *Nature* 138:1082-4
- 1947. Report on a visit to the site of the Eyasi skull, found by Dr. Kohl-Larsen. *Journal of East Africa Natural History Society* 19:40-3
- 1948. Fossil and sub-fossil Hominoidea in East Africa. In *The Robert Broom Commemorative volume*, edited by A.L. Du Toit, pp. 165-70. Cape Town: The Royal Society of South Africa
- 1952. The age of the Eyasi skull. In *Proceedings of the Pan-African congress on prehistory, 1947*, edited by L.S.B. Leakey. New York: Philosophical Library
- 1972. *Homo sapiens* in the Middle Pleistocene and the evidence of *Homo sapiens*' evolution. In *The origin of Homo sapiens*, edited by F. Bordes, pp. 25-9. Paris: UNESCO

Bibliography

Leakey, L.S.B. and M.D. Leakey. 1964. Recent discoveries of fossil hominids in Tanganyika at Olduvai and near Lake Natron. *Nature* 202:5-6

Leakey, L.S.B., P.V. Tobias, and J.R. Napier. 1964. A new species of the genus *Homo* from Olduvai Gorge. *Nature* 202:7-9

Leakey, M.D. 1971. Discovery of postcranial remains of *Homo erectus* and associated artefacts in Bed IV at Olduvai Gorge, Tanzania. *Nature* 232:380-3

— 1975. Cultural patterns in the Olduvai sequence. In *After the australopithecines: stratigraphy, ecology and culture change in the Middle Pleistocene*, edited by K.W. Butzer and G.Ll. Isaac, pp. 277-93. The Hague and Paris: Mouton

Leakey, M.D., R.L. Hay, G.H. Curtis, R.E. Drake, M.K. Jackes, and T.D. White. 1976. Fossil hominids from the Laetolil beds. *Nature* 262:460-6.

Leakey, M.G. and R.E.F. Leakey, editors. 1978. *Koobi Fora research project. Vol. 1: The fossil hominids and an introduction to their context 1968-1974.* London: Oxford University Press

Leakey, R.E.F. 1970. Fauna and artefacts from a new Plio-Pleistocene locality near Lake Rudolf in Kenya. *Nature* 226:223-4

— 1971. Further evidence of Lower Pleistocene hominids from East Rudolf, North Kenya. *Nature* 231:241-5

— 1972. Further evidence of Lower Pleistocene hominids from East Rudolf, North Kenya, 1971. *Nature* 237:264-9

— 1973a. Further evidence of Lower Pleistocene hominids from East Rudolf, North Kenya, 1972. *Nature* 242:170-3

— 1973b. Evidence for an advanced Plio-Pleistocene hominid from East Rudolf, Kenya. *Nature* 242:447-50

— 1974. Further evidence of Lower Pleistocene hominids from East Rudolf, North Kenya, 1973. *Nature* 248:653-6

— 1976a. New hominid fossils from the Koobi Fora formation in northern Kenya. *Nature* 261:574-6

— 1976b. Hominids in Africa. *American Scientist* 64:174-8

Leakey, R.E.F., M.G. Leakey, and A.K. Behrensmeyer. 1978. The hominid catalogue. In *Koobi Fora research project. Vol. 1: The fossil hominids and an introduction to their context 1968-1974*, edited by M.G. Leakey and R.E.F. Leakey. London: Oxford University Press

Leakey, R.E.F., J.M. Mungai, and A.C. Walker. 1971. New australopithecines from East Rudolf, Kenya. *American Journal of Physical Anthropology* 35:175-86

— 1972. New australopithecines from East Rudolf, Kenya (II). *American Journal of Physical Anthropology* 36:235-52

Leakey, R.E.F., M.D. Rose, and A.C. Walker. 1978. New *Homo erectus* specimens from East Turkana, Kenya. *American Journal of Physical Anthropology* (in press)

Leakey, R.E.F. and A.C. Walker. 1973. New australopithecines from East Rudolf, Kenya (III). *American Journal of Physical Anthropology* 39:205-22

— 1976. *Australopithecus, Homo erectus* and the single-species hypothesis. *Nature* 261:572-4

Leakey, R.E.F. and B.A. Wood. 1974a. New evidence of the genus *Homo* from East Rudolf, Kenya (IV). *American Journal of Physical Anthropology* 41:237-44

— 1974b. A hominid mandible from East Rudolf, Kenya. *American Journal of Physical Anthropology* 41:245-50

Le Gros Clark, W.E. 1955. *The fossil evidence for human evolution: an introduction to the study of paleoanthropology.* Chicago: University of Chicago Press

— 1964. *The fossil evidence for human evolution: an introduction to the study of paleoanthropology.* 2nd ed. Chicago: University of Chicago Press

Lieberman, P., E.S. Crelin, and D.H. Klatt. 1972. Phonetic ability and related anatomy of the newborn and adult human, Neanderthal man, and the chimpanzee. *American Anthropologist* 74:287-307

Limbrey, S. 1975. China. In *Catalogue of fossil hominids, part 3: Americas, Asia, and Australasia*, edited by K.P. Oakley, B G. Campbell, and T.I. Molleson, pp. 49-87. London: British Museum (Natural History)

Lisowski, F.P., G.H. Albrecht, and C.E. Oxnard. 1976. African fossil tali: further multivariate morphometric studies. *American Journal of Physical Anthropology* 45:5-18

de Lumley, H. 1966. Les fouilles de Terra Amata à Nice. Premiers résultats. *Bulletin du Musée d'anthropologie préhistorique de Monaco* 13:29-51

— 1971. Découverte de l'homme de l'Arago. *Le Courrier du Centre national de la recherche scientifique* 2(oct.):16-20

de Lumley, H. and M.-A. de Lumley. 1971. Découvertes de restes humains anténéandertaliens datés du début du Riss à la Caune de l'Arago (Tautavel Pyrénées Orientales). *Comptes-rendus de l'Academie des sciences, Paris*, Série D, Sciences Naturelles 272:1739-42

— 1975. Les hominiens quaternaires en Europe: mise au point des connaissances actuelles. *Centre national de la recherche scientifique, colloques internationaux, problèmes actuels de paléontologie (evolution des vertébrés)* 218:903-9. Paris

de Lumley, M.-A. 1970. Le pariétal humain anténéandertalien de Cova Negra (Jativa-Espagne). *Comptes-rendus de l'Academie des sciences, Paris*, Série D, Sciences Naturelles 270:39-41

- 1971. La mandibula de Banolas. *Ampurias* 33-34:1-91. Barcelona
- 1972. L'os iliaque anténéandertalien de la Grotte du Prince (Grimaldi, Ligurie italienne). *Bulletin du Musée d'anthropologie préhistorique de Monaco* 18:89-112
- 1973. Anténéandertaliens et néandertaliens du bassin Méditerranéen Occidental. *Etudes quaternaires, mémoire 2*
- 1975a. Ante-neanderthals of western Europe. In *Paleoanthropology: morphology and paleoecology*, edited by R.H. Tuttle, pp. 381-7. The Hague and Paris: Mouton
- 1975b. Meningeal vascularization of preneanderthalien man. In *Advances in cerebral angiography*, edited by G. Salamon, pp. 115-21. Berlin, Heidelberg, New York: Springer-Verlag
- 1976. Les anténéandertaliens dans le sud. In *La préhistoire française, I: les civilisations paléolithiques et mésolithiques*, edited by H. de Lumley, pp. 547-60. Paris: Centre national de la recherche scientifique

de Lumley, M.-A. and J. Piveteau. 1969. Les restes humains de la grotte du Lazaret (Nice, Alpes-Maritimes). *Mémoires de la Société préhistorique française* 7:223-32

Mania, D. 1973. Paläoökologie, faunenentwicklung und stratigraphie des eiszeitalters im mittleren Elbe-Saalegebiet auf Grund von molluskengesellschaften. *Geologie*, Beiheft 78 / 79
- 1974. Bilzingsleben, Kr. Artern — eine altpaläolithische travertinfundstelle im nördlichen mitteleuropa (vorbericht). *Zeitschrift für Archäologie* 8:157-73
- 1975. Bilzingsleben (Thüringen): eine neue altpaläolithische fundstelle mit knochenresten des *Homo erectus*. *Archaeologisches Korrespondenzblatt* 5:263-72
- 1976. Altpaläolithischer rastplatz mit hominidenresten aus dem mittelpleistozänen travertinkomplex von Bilzingsleben (DDR). 9e Congrés de l'Union internationale des sciences préhistoriques et protohistoriques, Nice, Colloque 9
- 1977. Die altpaläolithische travertinfundstelle von Bilzingsleben, Kr. Artern. *Ethnographisch-Archaeologische Zeitschrift* 18:5-24

Mania, D. and G.A. Cubuk. 1977. Altpaläolithische knochenspitzen vom typ Bilzingsleben-Hélin. *Zeitschrift für Archäologie* 11:87-92

Mania, D. and H. Grimm. 1974. Bilzingsleben, Kr. Artern — eine paläoökologisch aufschlussreiche fundstelle des altpaläolithikums mit hominiden-fund. *Biologische Rundschau* 12:361-4

Mania, D., H. Grimm, and E. Vlček. 1976. Ein weiterer hominiden-fund aus dem mittelpleistozänen travertinkomplex bei Bilzingsleben, Kr. Artern. *Zeitschrift für Archäologie* 10:241-9

Mania, D. and D.H. Mai. 1969. Warmzeitliche mollusken und pflanzenreste aus dem mittelpleistozän des geiseltals (südlich von Halle). *Geologie* 18:674-90

Mania, D. and E. Vlček. 1977. Altpaläolithische funde mit *Homo erectus* von Bilzingsleben (DDR). *Archeologické rozhledy* 29:603-16, 715-18

Marçais, J. 1934. Découverte de restes humains fossiles dans les grès quaternaires de Rabat (Maroc). *L'Anthropologie* 44:579-83

Martin, R. 1928. *Lehrbuch der anthropologie. II: Kraniologie, osteologie*. Jena: Gustav Fischer

Matthew, W.D. 1915. Climate and evolution. *Annals of the New York Academy of Sciences* 24:171-318

Maurer, G. 1968. Les montagnes du Rif central; étude géomorphologique. *Travaux de l'Institut scientifique chérifien, Rabat, série géologie et géographie physique* 14

Mayr, E. 1951. Taxonomic categories in fossil hominids. *Cold Spring Harbor Symposia on Quantitative Biology* 15:109-18

— 1963. The taxonomic evaluation of fossil hominids. In *Classification and human evolution*, edited by S.L. Washburn, pp. 332-46. Chicago: Aldine

McDougall, I., R. Maier, P. Sutherland-Hawkes, and A.J.W. Gleadow. 1980. K-Ar age estimate for the KBS tuff, East Turkana, Kenya. *Nature* 284:230-4

McHenry, H.M. 1975. Review of 'Uniqueness and diversity in human evolution; morphometric studies of australopithecines,' by C.E. Oxnard. *Science* 189:988

Mturi, A.A. 1976. New hominid from Lake Ndutu, Tanzania. *Nature* 262:484-5

Oakley, K.P. 1968. *Frameworks for dating fossil man*. 2nd ed. Chicago: Aldine

— 1975. Decorative and symbolic uses of vertebrate fossils. *Pitt Rivers Museum Occasional Papers on Technology* 12. Oxford: University Press

Oakley, K.P., B.G. Campbell, and T.I. Molleson, editors. 1975. *Catalogue of fossil hominids, part 3: Americas, Asia, Australasia*. London: British Museum (Natural History)

— 1977. *Catalogue of fossil hominids, part 1: Africa*. Second edition. London: British Museum (Natural History)

Olivier, G. and H. Tissier. 1975. Determination of cranial capacity in fossil men. *American Journal of Physical Anthropology* 43:353-62

Olson, T.R. 1978. Hominid phylogenetics and the existence of *Homo* in Member I of the Swartkrans Formation, South Africa. *Journal of Human Evolution* 7:159-78

Oppenoorth, W.F.F. 1932. De vondst van palaeolithische menschlijke schedels op Java. *De Mijninggenieur* 6:106-15

— 1933. The Java find of Neanderthal man. De vondst van palaeolithische menschelijke schedels op Java. *American Journal of Physical Anthropology* 17:240-4

Oxnard, C.E. 1975. *Uniqueness and diversity in human evolution: morphometric studies of australopithecines.* Chicago and London: University of Chicago Press

Oyen, O.J. and A.C. Walker. 1977. Stereometric craniometry. *American Journal of Physical Anthropology* 46:177-82

Patterson, B. and W.W. Howells. 1967. Hominid humeral fragment from early Pleistocene of northwestern Kenya. *Science* 156:64-6

Penrose, L.S. 1954. Distance, size and shape. *Annals of Eugenics* 18:337-43

Pilbeam, D. 1972. *The ascent of man: an introduction to human evolution.* New York: Macmillan

— 1975. Middle Pleistocene hominids. In *After the australopithecines: stratigraphy, ecology and culture change in the Middle Pleistocene,* edited by K.W. Butzer and G.Ll. Isaac, pp. 809-56. The Hague and Paris: Mouton

Piveteau, J. 1967. Un pariétal humain de la grotte du Lazaret (Alpes-Maritimes). *Annales de paléontologie (vertébrés)* 53:167-99

— 1973. *Origine et destinée de l'homme.* Paris: Masson

— 1976a. Les anté-néandertaliens du sud-ouest. In *La préhistoire française, I: les civilisations paléolithiques et mésolithiques,* edited by H. de Lumley, pp. 561-6. Paris: Centre national de la recherche scientifique

— 1976b. Les anté-néandertaliens du sud-ouest de la France. In *Le peuplement anténéandertalien de l'Europe,* edited by A. Thoma, pp. 29-30. 9e Congrés de l'Union internationale des sciences préhistoriques et protohistoriques, Nice, Colloque 9

Protsch, R. 1975. The absolute dating of Upper Pleistocene subsaharan fossil hominids and their place in human evolution. *Journal of Human Evolution* 4:297-322

— 1976. The position of the Eyasi and Garusi hominids in East Africa. In *Les plus anciens hominides,* edited by P.V. Tobias and Y. Coppens, pp. 207-38. 9e Congrés de l'Union internationale des sciences préhistoriques et protohistoriques, Nice, Colloque 9

— 1981. The palaeoanthropological finds of the Plicene and Pleistocene: part I, monograph Garusi; part II, monograph Eyasi. Volume V of *Die Archäologischen und Anthropologischen Ergebnisse der Kohl-Larsen Expeditionen in Nordtanzania 1933-1939,* edited by H. Müller-Beck. Stuttgart, Berlin, Köln, Mainz: Archaeologica Venatoria W. Kohlhammer

Raynal, R. 1961. *Plaines et piedmonts du bassin de la Moulouya (Maroc oriental): étude géomorphologique.* Rabat: Imprimeries françaises et marocaines

Read, B.E. 1934. Chinese materia medica: dragons and snakes. *Peking Natural History Bulletin* 8(4):1-66

Reck, H. and L. Kohl-Larsen. 1936. Erster uberblick über die jungdiluvialen tier- und menschenfunde Dr. Kohl-Larsen's im nordöstlichen Teil des Njarasa-grabens (Ostafrika). *Geologische Rundschau* 27:421-41

Reeve, W.H. 1946. Geological report on the site of Dr. Kohl-Larsen's discovery of a fossil human skull, Lake Eyasi, Tanganyika Territory. *East Africa Natural History Society Journal* 19:44-50

Remane, A. 1951. Die zähne des *Meganthropus africanus*. *Zeitschrift für Morphologie und Anthropologie* 42:311-29

— 1954. Structure and relationships of *Meganthropus africanus*. *American Journal of Physical Anthropology* 12:123-6

Rhoads, J.G and E. Trinkaus. 1977. Morphometrics of the Neandertal talus. *American Journal of Physical Anthropology* 46:29-43

Rightmire, G.P. 1976. Relationships of Middle and Upper Pleistocene hominids from sub-Saharan Africa. *Nature* 260:238-40

— 1979. Cranial remains of *Homo erectus* from Beds II and IV, Olduvai Gorge, Tanzania. *American Journal of Physical Anthropology* 51:99-115

— 1980. Middle Pleistocene hominids from Olduvai Gorge, northern Tanzania. *American Journal of Physical Anthropology* 53:225-41

Robinson, J.T. 1953. *Meganthropus*, australopithecines and hominids. *American Journal of Physical Anthropology* 11:1-38

— 1961. The australopithecines and their bearing on the origin of man and of stone-tool making. *South African Journal of Science* 57:3-13

— 1968. The origin and adaptive radiation of the australopithecines. In *Evolution and hominisation*, edited by G. Kurth, pp. 150-75. Stuttgart: Gustav Fischer

— 1972a. *Early hominid posture and locomotion*. Chicago and London: University of Chicago Press

— 1972b. The bearing of East Rudolf fossils on early hominid systematics. *Nature* 240:239-40

Roche, J. 1959. Nouvelle datation de l'epipaléolithique marocain par la méthode du carbone 14. *Comptes-rendus de l'Academie des sciences, Paris* 249:729-30

Roche, J. and J.-P. Texier. 1976. Découverte de restes humains dans un niveau artérien superieur de la grotte des contrabandiers à Témara (Maroc). *Comptes-rendus de l'Academie des sciences, Paris,* Série D, Sciences Naturelles 282:45-7

Saban, R. 1972. Les hommes fossiles du Maghreb. *L'Ouest médical* 25(23):2443-58

— 1975. Les restes humains de Rabat (Kébibat). *Annales de paléontologie (vertébrés)* 61:153-207. Paris

Santa Luca, A.P. 1976. Solo man and Neandertal man: morphological and metrical comparisons. *American Journal of Physical Anthropology* 44:203/(abstract)
- 1977. A comparative study of the Ngandong fossil hominids. PhD dissertation, Harvard University, Cambridge
- 1978. A re-examination of presumed Neandertal-like fossils. *Journal of Human Evolution* 7:619-36

Sartono, S. 1968. Early man in Java: Pithecanthropus skull 7, a male specimen of *Pithecanthropus erectus*. *Proceedings Koninklijke Nederlandse Akademie van Wetenschappen*, Series B, Physical Sciences 71:396-422
- 1975. Implications arising from Pithecanthropus VIII. In *Paleoanthropology: morphology and paleoecology*, edited by R.H. Tuttle, pp. 327-60. The Hague: Mouton

Sausse, F. 1975. La mandibule atlanthropienne de la carrière Thomas I (Casablanca). *L'Anthropologie* 79:81-112

Schlosser, M. 1903. Die fossielen säugetiere Chinas. *Abhandlungen der Bayerische Akademie der Wissenschaften* 22:1-221

Senyürek, M. 1955. A note on the teeth of *Meganthropus africanus* Weinert from Tanganyika Territory. *Belleten* 19(73):1-57

Shackleton, N.J. and N.D. Opdyke. 1973. Oxygen isotope and paleomagnetic stratigraphy of equatorial Pacific core V28-238: oxygen isotope temperatures and ice volumes on a 10^5 year and 10^6 year scale. *Quaternary Research* 3:39-55

Shapiro, H.L. 1974. *Peking man*. New York: Simon and Schuster

Siedner, G. and A. Horowitz. 1974. Radiometric ages of late Cainozoic basalts from northern Israel: chronostratigraphic implications. *Nature* 250:23-6

Simons, E.L., D. Pilbeam, and P.C. Ettel. 1969. Controversial taxonomy of fossil hominids. *Science* 166:258-9

Simpson, G.G. 1945. The principles of classification and a classification of mammals. *Bulletin of the American Museum of Natural History* 85:1-350
- 1961. *Principles of animal taxonomy*. New York: Columbia University Press

Sivak, J. 1977. Les pollens de Cathaya et de Tsuga du gisement pliocène du lac Ichkeul (Tunisie). *Recherches françaises sur le quaternaire hors de France*. Supplément au Bulletin de l'AFEQ, Congrés INQUA de Birmingham à paraître.

Smith, G.E. 1913. The Piltdown skull and brain cast. *Nature* 92:131, 267-8, 318-9
- 1934a. Davidson Black 1884-1934. *Obituary Notices of Fellows of the Royal Society, 1932-1935* 3:361-5
- 1934b. Obituary: Prof. Davidson Black, FRS. *Nature* 133:521-2

Stearns, C.E. and D.L. Thurber. 1965. Th^{230}/U^{234} dates of late Pleistocene marine fossils from the Mediterranean and Moroccan littorals. *Quaternaria* 7:29-42

Stevenson, P.H. 1934. In *Davidson Black, 1884-1934, in memoriam*, pp. 1-15. Geological Society of China, Peking Society of Natural History

Stewart, T.D. 1970. The evolution of man in Asia as seen in the lower jaw. *Proceedings 8th International Congress of Anthropological and Ethnological Sciences* 1:263-6

Stringer, C.B. 1974. Population relationships of later Pleistocene hominids: a multivariate study of available crania. *Journal of Archaeological Science* 1:317-42

Stringer, C.B., F.C. Howell, and J.K. Melentis. 1979. The significance of the fossil hominid skull from Petralona, Greece. *Journal of Archaeological Science* 6:235-53

Taieb, M., D.C. Johanson, Y. Coppens, and J.L. Aronson. 1976. Geological and palaeontological background of Hadar hominid site, Afar, Ethiopia. *Nature* 260:289-93

Tchernov, E. 1968. *Succession of rodent faunas during the Upper Pleistocene of Israel*. Hamburg: Parey

— 1975. Rodent faunas and environmental changes in the Pleistocene of Israel. In *Rodents in desert environments*, edited by I. Prakash and P.K. Ghosh, pp. 331-62. The Hague: DR. W. Junk

Thoma, A. 1966. L'occipital de l'homme mindélien de Vértesszöllös. *L'Anthropologie* 70:495-538

— 1967. Human teeth from the Lower Palaeolithic of Hungary. *Zeitschrift für Morphologie und Anthropologie* 58:152-80

— 1969. Biometrische studie über das occipitale von Vértesszöllös. *Zeitschrift für Morphologie und Anthropologie* 60:229-41

— 1976a. *Le peuplement anténéandertalien de l'Europe*. 9e Congrés de l'Union internationale des sciences préhistoriques et protohistoriques, Nice, Colloque 9

— 1976b. Le peuplement anténéandertalien d'Europe dans le contexte paléo-anthropologique de l'ancien monde. In *Le peuplement anténéandertalien de l'Europe*, edited by A. Thoma, pp. 7-16. 9e Congrés de l'Union internationale des sciences prehistoriques et protohistoriques, Nice, Colloque 9 d'études démographiques

— 1976c. Corrélations génétiques entre populations de chausseurs. In *L'étude des isolats; espoirs et limites*, edited by A. Jacquard. Paris: Institut national d'études démographiques

Ting, V.K. 1934. In *Davidson Black, 1884-1934, in memoriam*, pp. 16-19. Geological Society of China, Peking Society of Natural History

Tobias, P.V. 1967. *Olduvai Gorge, volume 2: the cranium and maxillary dentition of Australopithecus (Zinjanthropus) boisei.* Cambridge: Cambridge University Press

— 1968a. Middle and early Upper Pleistocene members of the genus *Homo* in Africa. In *Evolution and hominisation*, edited by G. Kurth, pp. 1976-94. Stuttgart: Gustav Fischer

— 1968b. The taxonomy and phylogeny of the australopithecines. In *Taxonomy and phylogeny of old world primates with references to the origin of man*, edited by B. Chiarelli, pp. 277-315. Torino: Rosenberg and Sellier

— 1973. Darwin's prediction and the African emergence of the genus *Homo*. *Academia Nazionale dei Lincei* 182:63-85

Tobias, P.V. and G.H.R. von Koenigswald. 1964. A comparison between the Olduvai hominines and those of Java, and some implications for hominid phylogeny. *Nature* 204:515-18

Toepfer, V. 1960. Das letztinterglaziale mikrolithische paläolithikum von Bilzingsleben, Kr. Artern. *Ausgrabungen und Funde* 5:7-11

— 1963. Bemerkungen zum geologischen alter und zu den Kernsteinen paläolithischer kulturen aus dem Eem-interglazial in mitteldeutschland. *Alt-Thüringen* 6:42-56

— 1970. Stratigraphie und okologie des paläolithikums. Periglazial-löss-paläolithikum im jungpleistozän der Deutschen Demokratischen Republik. *Petermanns Geographische Mitteilungen Ergänzungsheft* 274:329-422

Trinkaus, E. 1973. A reconsideration of the Fontéchevade fossils. *American Journal of Physical Anthropology* 39:25-35

Twiesselmann, F. 1941. Méthode pour l'evaluation de l'épaisseur des parois crâniennes. *Bulletins du Musée royal d'histoire naturelle de belgique* 17(48):1-33

Vallois, H.V. 1945. L'homme fossile de Rabat. *Comptes-rendus de l'Academie des sciences, Paris* 221:669-71

— 1955. La mandibule humaine pré-moustérienne de Montmaurin. *Comptes-rendus de l'Academie des sciences, Paris* 240:1577-9

— 1958. La grotte de Fontéchevade: partie 2, anthropologie. *Archives de l'Institut de paléontologie humaine, mémoire* 29

— 1959. L'homme de Rabat. *Bulletin d'archéologie marocaine* 3: 87-91

Vallois, H.V. and J. Roche. 1958. La mandibule acheuléenne de Témara, Maroc. *Comptes-rendus de l'Academie des sciences, Paris* 246:3113-16

Van Campo, E. 1977. Une flore sporopollinique du gisement pliocëne du lac Ichkeul (Tunisie). *Recherches françaises sur le quaternaire hors de France.* Supplément au Bulletin de l'AFEQ, Congrés INQUA de Birmingham à paraître

Van der Meulen, A.J., and W.H. Zagwijn. 1974. *Microtus (Allophaiomys) pliocaenicus* from the Lower Pleistocene near Brielle, The Netherlands. *Scripta Geologica* 21:1-12. Leiden

Van Monfrans, H.M. 1971. Palaeomagnetic dating in the North Sea basin. Unpublished dissertation, Amsterdam University

Vlček, E. 1978. A new discovery of *Homo erectus* in central Europe. *Journal of Human Evolution* 7:239-51

— 1979. Die mittelpleistozänen hominidenreste von Bilzingsleben. Veröffentlichungen des Landesmuseums für Vorgeschichte in Halle (Salle), Berlin (in press)

Vlček, E. and D. Mania. 1977. Ein neuer fund von *Homo erectus* in Europa Bilzingsleben (DDR). *Anthropologie* (Brno) 15:159-69

Vondra, C.F. and B.E. Bowen. 1976. Plio-Pleistocene deposits and environments, East Rudolf, Kenya. In *Earliest man and environments in the Lake Rudolf basin: stratigraphy, paleoecology, and evolution*, edited by Y. Coppens, F.C. Howell, G.Ll. Isaac, and R.E.F. Leakey, pp. 79-93. Chicago and London: University of Chicago Press

— 1978. Stratigraphy, sedimentary facies and paleoenvironments, East Lake Turkana, Kenya. In *Geological background to fossil man*, edited by W.W. Bishop, pp. 395-414. Edinburgh: Scottish Academic Press

Vondra, C.F., G.D. Johnson, B.E. Bowen, and A.K. Behrensmeyer. 1971. Preliminary stratigraphical studies of the East Rudolf basin, Kenya. *Nature* 231:245-8

Walker, A. 1976. Remains attributable to *Australopithecus* in the East Rudolf succession. In *Earliest man and environments in the Lake Rudolf basin: stratigraphy, paleoecology, and evolution*, edited by Y. Coppens, F.C. Howell, G.Ll. Isaac, and R.E.F. Leakey, pp. 484-9. Chicago and London: University of Chicago Press

Walker, A. and R.E.F. Leakey. 1978. The hominids of East Turkana. *Scientific American* 239(2):54-66

Weidenreich, F. 1935. The *Sinanthropus* population of Choukoutien (locality1) with a preliminary report on new discoveries. *Bulletin of the Geological Society of China* 14:427-68

— 1936. The mandibles of *Sinanthropus pekinensis*: a comparative study. *Palaeontologia Sinica*, Series D, 7(3)

— 1937a. The dentition of *Sinanthropus pekinensis*: a comparative odontography of the hominids. *Palaeontologia Sinica*, New Series D, 1

- 1937b. The new discoveries of *Sinanthropus pekinensis* and their bearing on the *Sinanthropus* and *Pithecanthropus* problems. *Bulletin of the Geological Society of China* 16:439-70
- 1937c. The relation of *Sinanthropus pekinensis* to *Pithecanthropus, Javanthropus*, and Rhodesian Man. *Journal of the Royal Anthropological Institute* 67:51-65
- 1940. Some problems dealing with ancient man. *American Anthropologist* 42:375-83
- 1941. The extremity bones of *Sinanthropus pekinensis*. *Palaeontologia Sinica*, New Series D, 5
- 1943. The skull of *Sinanthropus pekinensis*: a comparative study on a primitive hominid skull. *Palaeontologia Sinica*, New Series D, 10
- 1946. *Apes, giants and man*. Chicago: University of Chicago Press
- 1948. The duration of life of fossil man in China and the pathological lesions found in his skeleton. In *Anthropological Papers of Franz Weidenreich 1939-1948: a memorial volume*, edited by S.L. Washburn and D. Wolffson, pp. 194-204. New York: Viking Fund
- 1951. Morphology of Solo man. *Anthropological Papers of the American Museum of Natural History* 43:205-90

Weiner, J.S. and B.G. Campbell. 1964. The taxonomic status of the Swanscombe skull. In *The Swanscombe skull: a survey of research on a Pleistocene site*, edited by C.D. Ovey, pp. 175-209. *Royal Anthropological Institute Occasional Paper* 20

Weinert, H. 1937. Hominidae (Paläozoologie). *Fortschritte der Paläontologie* 1:337-44
- 1938a. Der neue affenmensch '*Africanthropus.*' *Germania* 1:21-4
- 1938b. *Afrikanthropus*: der erste affenmenschen-fund aus dem quatär Deutsch-Ostafrikas. *Quatär* 1:177-9
- 1938c. *Africanthropus* (*Pithecanthropus* stufe). In *Enstehung der menschenrassen*, pp. 33-40. Stuttgart: F. Enke
- 1938d. Der erste afrikanische affenmensch, '*Africanthropus njarasensis.*' *Der Biologe* 7(4):125-9
- 1940. *Africanthropus*, der neue affenmenschenfund vom Njarasa-See in Ostafrika. *Zeitschrift für Morphologie und Anthropologie* 38:18-24
- 1950. Uber die neuen vor- und frühmenschenfunde aus Afrika, Java, China und Frankreich. *Zeitschrift für Morphologie und Anthropologie* 42:113-48
- 1952. Uber die vielgestaltigkeit der summopromaten vor der menschwerdung. III: *Australopithecus und Paranthropus*. *Zeitschrift für Morphologie und Anthropologie* 43:311-30

Weinert, H., W. Bauermeister, and A. Remane. 1940. *Africanthropus njarasensis.* Beschreibung und phylethische einordnung des ersten affenmenschen aus Ostafrika. *Zeitschrift für Morphologie und Anthropologie* 38:252-308

Wells, L.H. 1947. A note on the maxillary fragment from the Broken Hill cave. *Journal of the Royal Anthropological Institute* 77(1):11-12

White, T.D. and J.M. Harris. 1977. Suid evolution and correlation of African hominid localities. *Science* 198:13-21

Wiegers, F. 1928. *Diluviale vorgeschichte des menschen.* Stuttgart: F. Enke

— 1940. Das geologische alter der altsteinzeitlichen kulturen von Wangen a.d. Unstrut und Bilzingsleben a.d. Wipper. *Prähistorische Zeitschrift* 30/31:331-6

Wolpoff, M.H. 1971. Competitive exclusion among Lower Pleistocene hominids: the single species hypothesis. *Man* 6:601-14

— 1971. Vertesszöllös and the presapiens theory. *American Journal of Physical Anthropology* 35:209-15

— 1976. Some aspects of the evolution of early hominid sexual dimorphism. *Current Anthropology* 17:579-606

Wong-Staal, F., D. Gillespie, and R.C. Gallo. 1976. Proviral sequences of baboon endogenous type C RNA virus in the DNA of human leukaemic tissues. *Nature* 262:190-5

Woo, J.K. 1962. The mandibles and dentition of *Gigantopithecus. Palaeontologia Sinica*, New Series D, 11:1-94

— 1964. Mandible of Sinanthropus lantianensis. *Current Anthropology* 5:98-101

— 1966. The skull of Lantian Man. *Current Anthropology* 7:83-6

Woo, J.K. and R.C. Peng. 1959. Fossil human skull of early palaeoanthropic stage found at Mapa, Shaoquan, Kwangtung province. *Vertebrata Palasiatica* 3:176-82

Wood, B.A. 1976. Remains attributable to *Homo* in the East Rudolf succession. In *Earliest man and environments in the Lake Rudolf basin: stratigraphy, paleoecology, and evolution,* edited by Y. Coppens, F.C. Howell, G.Ll. Isaac, and R.E.F. Leakey, pp. 490-506. Chicago and London: University of Chicago Press

Zdansky, O. 1928. Die säugetiere der quartärfauna von Chou-K'ou-Tien. *Palaeontologia Sinica*, Series C, 5:1-146

— 1952. A new tooth of *Sinanthropus pekinensis* Black. *Acta Zoologica* 33:189-91

Index

Academia Sinica, Peking 44
Acheulean 76, 163, 191
Afar, Ethiopia 80, 84, 213, 233
Africanthropus njarasensis 220
allopatric speciation 71, 81, 82
Ambrona 119
American Museum of Natural History, New York 5, 33, 43-8, 52, 53
amino acid racemization 8, 10, 134, 155, 221, 224, 230, 232
Amud 76
Andersson, J.G. 22, 32
Andrews, R.C. 64
Anténéanderthal (Anteneanderthal, Preneanderthal) 8, 11, 75, 76, 118, 119, 121-3, 126-8, 130-2, 154, 157
l'Arago 75, 112, 118, 119, 120-3, 126, 128, 130, 143, 144, 154, 156, 231
Archanthropine 65, 67, 68, 72, 116, 118, 119, 124, 126, 132
Atapuerca 120, 130
Atlanthropus 111, 121, 128; *Atlanthropus mauritanicus*, 231
Australoids (Australians) 92, 113, 114, 223
australopithecines 66, 67, 78-83, 98, 154, 223, 224, 229, 230, 233

Australopithecus 35, 66, 78, 80, 83, 84, 88, 201, 203, 209, 211-13
Australopithecus afarensis 84, 85, 233
Australopithecus africanus 21, 66, 79, 210, 212, 213, 224, 229, 231, 233
Australopithecus boisei 209, 212
Australopithecus robustus 66, 209, 211, 212, 224, 233

Bañolas 120, 130
Bantus 223
Barrel, L. 127
Biache 112, 231
Bilzingsleben ix, 4, 7, 8, 11, 111, 133-7, 143, 144, 146, 147, 149-51, 155, 156, 232, 233
biospecies 70, 82
Black, Davidson vii-ix, 3-6, 10, 12-15, 21-39, 32-4, 42, 43, 52, 64, 101, 133
Black III, Davidson 4, 12
Bohlin, B. 22, 32, 41, 64
Bordes, F. 128
Bowers, J.Z. 24
British Museum of Natural History 43, 44, 46-8
Broken Hill 64, 65, 68, 72, 76, 77, 144, 221, 232

266 Index

Broom, R. 21
Brunhes–Matuyama 172, 191, 228

Canada vii-x, 12, 14, 23, 228
Canadian Association for Physical Anthropology/l'Association pour l'Anthropologie Physique du Canada viii, 3; newletter, 228
cannibalism 7, 97-101, 221
Carabelli's pit 37
Cave of Hearths 76
Clactonian technique 145
Chesowanja 209
Chari/Karari tuffs 197
Chenchiawo locality 228
Chih, H. Ch'en 43
China 5, 22, 26-9, 33-5, 37, 39, 43, 48, 73, 74, 80, 82, 85, 101, 143, 209, 212, 227, 228, 233; People's Republic of 3
Chinese drugstores 28, 29, 34, 35, 39; 'Dragon' teeth from 27, 28, 34, 35
Chou Kou Tien: site vii, viii, 4, 6, 13, 15, 22, 31-4, 37, 42-7, 49, 56, 57, 60, 61, 64, 67, 73, 82, 83, 133, 153; Locality 1 41, 45, 48, 50-3, 58, 60, 73, 74, 227, 228, 230; Lower Cave 52, 55; Upper Cave 42, 45, 49, 92
cingulum 37
chronocline 186
clades: in evolution 66, 68, 71, 72, 76, 77, 81, 84; cladistic 70, 71, 84, 194, 212, 213; cladogenetic 83; cladograms 213; as grades in evolution (see grades) 66-72, 74, 77, 83, 84
cline 209
Combier, J. 126

Coon, C. 68, 69, 71, 73, 75, 77
Cova Negra 75, 120, 128, 129, 130
Cowdry, E. 14, 24

Dart, R. 21
David, P. 128
Débénath, A. 106, 127, 128
de Chardin, Teilhard vii, 36
Delattre, A. 130
de Lumley, H. 119, 120, 128
de Lumley, M.-A. 154-6
Djetis faunal zone 72-4, 80, 229
DNA hybridization, for measuring genetic distance 80, 81
Dryopithecus, cusp pattern 105
Dubois, E. 6, 21, 32, 63, 64

East Africa 6, 78, 81-4, 143, 153, 189, 192, 212, 218, 222, 226, 230, 233
East Turkana (or East Rudolf, or Lake Turkana area) 9-11, 77, 79, 153, 189, 194, 211-14, 229-31
Ehringsdorf 143, 144
Elster glaciation 134, 145
Elster/Mindel interglacial 8, 134
Eyasi 10, 217-23, 226, 232

Field Museum of Natural History, Chicago: casts of *Sinanthropus* 43
Fenart, R. 130
fire, use of 119
fission track dating 194, 199, 230
Florisbad 76, 220
fluviatile deposits 195
Fontéchevade 75, 120, 127, 129, 232

Garusi 10, 217-19, 224-6, 232, 233
genus, criteria for defining 66

Gigantopithecus 5, 38, 39; *Gigantopithecus blacki* 5, 38, 39
grades (see clades); in evolution 66-72, 74, 77, 83, 84

Haberer, K.A. 28, 34-6
Haeckel, E. 63, 65, 66
Harvard Peabody Museum: casts of *Sinanthropus* 43, 48
Hemanthropus 5, 35
Hemmer, H. 156
Henri-Martin, G. 127
Hipparion 28, 29, 35
Holstein 8, 134, 151
Homo 8-10, 39, 66, 67, 69, 71, 73, 77-85, 111, 113, 143, 203, 211, 212, 226, 229, 230, 231, 233
Homo africanus 66, 78, 79, 229, 231, 233
Homo erectus viii, ix, 3-11, 60, 63, 64, 66-72, 74-84, 111, 112, 114, 132, 143, 144, 151, 153-7, 159, 177, 179, 181, 184-7, 189, 191, 192, 205, 209-14, 217, 218, 220-2, 226-35
Homo erectus bilzingslebensis 8, 11, 144
Homo erectus erectus 179, 184
Homo erectus javanicus 67
Homo erectus mauritanicus 177, 178, 181
Homo erectus pekinensis 42, 67, 184
Homo (erectus seu sapiens) palaeohungaricus 7, 11, 112
Homo ergaster 79
Homo habilis 21, 78, 79, 81, 82, 84, 205, 213, 229, 231, 233
Homo modjokertensis 229, 231
Homo neanderthalensis 67

Homo sapiens 6, 7, 9, 11, 36, 38, 49, 65-72, 74-7, 80, 83, 84, 88, 112, 114, 143, 144, 153, 186, 192, 224, 230, 232, 233
Homo sapiens erectus 114
Homo sapiens neanderthalensis 77
Homo sapiens rhodesiensis 10, 65, 77, 221, 222, 226, 232
Homo sapiens sapiens 92, 144, 185, 186
Homo soloensis 144, 228
Homo transvaalensis 66
Hood, D. 13, 24, 64
Howell, F.C. 82, 85
Hungarian Museum of Natural History, Budapest 106
Hupeh province 228
hybridization 77

Ikeda, J. 222
Ileret: region 197; tuff complex 202
Indonesia 5-7, 87, 89, 90, 101, 229
Institute of Vertebrate Paleontology and Paleoanthropology, Peking 43
Isaac, G. 193

Japan (Japanese): occupation in China 34; war in China 42; occupation in Java 33; invasion, Chinese fear of (1941) 48
Java: cranium 64; country 5, 6, 33-5, 63, 65, 72-4, 80-2, 85, 100, 143, 209, 211, 212, 223, 227, 228, 233; fossils 66, 67; Java Man 32, 65, 95; Javanese paleontology 7; Javanese taxa 7

Kabul Beds 89-91, 93, 229
Kanapoi 79, 81
Kappers A. 24

268 Index

KBS tuff 10, 195, 197, 198, 212, 215, 229
Kedungbrubus 89, 90
Kohl-Larsen 222; expedition to Tanzania 1934-40 10, 218, 219
Koobi Fora: sediments 193, 201; hominids 194, 200, 201; Research Project 194; region 195-7; formation 197, 201-3, 205, 207-9, 212, 229
Krapina 35, 127
Kretzoi, M. 112
Kromdraai 209, 230
Kungwangling locality 228

La Chaise caves 75, 106
la Chapelle-aux-Saints 127
lacustrine deposits 134, 138, 195
Laetolil 79, 219, 226, 233; Beds 10, 223, 224, 226; hominids 224-6, 233; River 223
La Ferrassie 125, 127
Lantian fossils 73, 227-9
La Quina 108
Lazaret: cave site 120, 127, 131; fossils 126, 130
Leakey, L.S.B. 21, 78, 219, 220, 222
Leakey, M. 78, 79, 224, 226
Leakey, R.E.F. 79, 193
Le Gros Clark, Sir W.E. 67, 69
l'Hortus 127
Lunel-Viel 119

Maghreb 9, 159-61, 163, 166, 168-70, 173-9, 184-7
Makapansgat 230
Mapa 76, 228
marine deposits 82
Masek Beds 191, 230
Matthew, W.D. 22

Mauer 7, 64, 71, 111, 118, 120, 121, 130, 131, 154, 155, 232, 233
Mayr, E. 66, 67, 69, 70
Meganthropus 35, 72, 211, 223; *Meganthropus africanus* 223
Melentis, J. 155
Mindelian 7, 105, 118, 119, 156
Mongoloids 92
Montmaurin 118-121, 130, 154
multivariate analyses 72, 75, 79-81, 232

Ndutu cranium 76, 222, 230, 233
Neanderthal 35, 61, 63-8, 71, 72, 74-7, 82, 83, 92, 93, 106, 112, 115, 118, 119, 121, 124-31, 143, 144, 154-6, 220, 221, 228, 229, 232
Neanthropine 65
neotony 186
Ngandong faunal zone 72, 73, 85, 89-91, 94, 229
nomenclature: problems 10, 11; correct procedures 71

Oberkassel fossils 110
Oldowan industry, developed 191
Olduvai 11, 68, 77, 78, 80, 192, 209, 213; Olduvai Gorge 4, 9, 153, 189, 201, 218, 219, 223, 229-31, 233
Omo: hominids 65, 75, 76; Omo Valley 77, 79
Orgnac 120, 126, 127
Osborne, H.F. 22, 25
Oxnard, C. 81

Palaeoanthropine 65
Palaeoanthropus njarasensis 221
paleoecology 139

paleomagnetic dating 90, 91, 194, 195, 197, 199, 228, 230; magnetic polarity 189
paleospecies 11, 70, 112
Paranthropus 5, 21, 66, 78, 81, 113, 209, 230, 233
Pei, W.C. vii, 22
Peking 14, 22-7, 29, 33, 34, 36, 43, 65, 68, 207, 211, 228; Peking Man vii, 10, 13, 21, 22, 25, 27, 29, 32-4, 36-8, 64-6, 74, 76, 92, 95, 97, 101, 227-9; Peking hominids 60; Peking Union Medical College 22, 24, 48
Perez, L. 204, 205
Perning 89, 229
Petralona 7, 111, 155, 156, 232, 233
phyletic gradualism 70, 71, 73, 77, 186
Piltdown Man 24
pithecanthropines 87, 89, 90, 93, 95, 98, 101, 114
Pithecanthropus 11, 32, 33, 64, 67, 89, 97, 98, 100, 111, 113, 121, 124, 220, 229
Pithecanthropus alalus 63
Pithecanthropus erectus 6, 7, 21, 63, 67, 72, 89, 92, 93, 96, 97, 229
Pithecanthropus lantianensis 94, 97
Pithecanthropus modjokertensis 6, 7, 73, 89, 91, 92, 94, 97
Pithecanthropus pekinensis 93-95, 101
Pithecanthropus robustus 72
Pithecanthropus (=Sinanthropus) lantianensis 7
Pithecanthropus (=Sinanthropus) pekinensis 7, 92
Pithecanthropus soloensis 6, 7, 72, 73, 89, 91-4, 96, 101

Piveteau, J. 128
platycephaly 222
Plesianthropus 21; *Plesianthropus transvaalensis* 224
polymorphism 8
Předmost 77
Prezletice 118
Prince, la grotte du 120, 127
Protsch, R. 134
Puchangan deposits 80, 90, 91
punctuated equilibria 70, 72, 84

Qafza 76

Rabat 112, 160, 162, 178, 179, 181, 183, 184, 187, 231-3
radiometric dating 97, 101, 197, 233; uranium-thorium 172, 232; potassium-argon 189, 194, 219, 230
Read, B.E. 24
Reck, H. 220, 222
Rhodesian Man 65, 76
Rift Valley 9, 10
Riss 106, 111, 118, 120, 126-8, 130, 228
riverine deposits 134
Robinson, J.T. 4, 79, 113

Saale 134
Saale/Riss 8, 134, 138
Saccopastore 112
Saldanha 65, 76, 77, 221, 232
Salé 9, 75, 160, 162, 178-87, 231-3
Sambungmachan 73, 89, 91, 94
Sangiran 33, 77, 85, 89-91, 93, 229
Schlosser, M. 29, 34, 35
sexing: of *Sinanthropus* 56-60; in general in fossil hominids 61

sexual dimorphism 186, 210; in dentition 56-8; in *Sinanthropus* 60; in extinct hominids 61
Shanidar 76
Sickenberg, O. 156
Sidi Abderrhamen 160, 162, 178, 179, 184, 231, 233
Simone, S. 127
Sinanthropus viii, 4-6, 14, 15, 33, 38, 43-5, 48, 49, 53, 59, 106, 111, 121, 124, 125, 128, 223, 224, 227
Sinanthropus officinalis 38
Sinanthropus pekinensis vii, 3, 4, 10, 13, 15, 22, 25, 27, 32, 42, 64, 67
single species hypothesis 209, 211
Skhul 76, 77, 110
Smith, G.E. 14, 24
Solo: Man 33, 67, 68, 72, 92-5, 97, 101, 129; fossils 65, 73, 74, 77, 232; River 94
South Africa 3, 35, 75, 78, 81, 83, 143, 212, 230, 233
Spengler, A. 133
Steinheim 65, 75, 112, 121, 124, 125, 143, 144, 154, 221, 231, 232
Sterkfontein 77, 210, 211, 213, 223, 230
Suard 120, 128
Swanscombe 65, 75, 106, 108-10, 112, 129, 143, 144, 154, 155, 187, 231, 232
Swartkrans 35, 77, 209, 213, 223, 224, 230, 231
Sweden 29, 31, 32, 41; Swedish Academy 29; Palaeontological Institute, Uppsala 44
sympatry 82, 97

Tabun 76
taphonomy 193, 200

Taschdjian, C. 43, 48
Telanthropus 67, 230, 231
Ternifine 9, 67, 71, 160, 162, 168, 172, 175, 177-9, 184-7, 192, 230, 231
Terra Amata 119
Teshik Tash 76
Thomas I deposits 162, 168, 169, 177-9
Thomas III deposits 160, 162, 178, 179, 181, 184
Ting, V.K. 14, 26
Tobias, P.V. 77, 106, 108
tool traditions: pebble-chopper 105; acheulean 111; African-type cleavers 112
Toronto 23, 28
Torralba 119
Torre in Pietra, Italy 119
travertine 134, 138, 151, 232
Trinil faunal zone 32, 63, 68, 72, 73, 80, 83, 89, 90, 92, 229
Trogontherium 148, 151
Tulu Bor tuff 197

Ubeidiya deposits 172
University of Pennsylvania Museum in Philadelphia: casts of *Sinanthropus* 43, 48
University of Toronto vii-ix, 3, 4, 12, 14, 23

Vallois, H.V. 129
Vergranne 119
Vértesszöllös 7, 11, 105, 106, 111, 118, 119, 143, 144, 155, 156, 187, 232, 233
Vines, Père Gonzalo 128
von Koenigswald, G.H.R. 45; casts of *Sinanthropus* in Frankfurt 43

von Zittel 28

Weichsel-Würm glaciation 138
Weidenreich, F. vii, 3, 6, 29, 33, 34,
    38, 43, 45, 48, 49, 52-60, 65, 67,
    68, 71, 73, 77, 92, 93, 113
Weinert, H. 217, 218, 220, 222
Wenner-Gren Foundation: casts of
    Sinanthropus 43, 45
Western Reserve University, Cleveland,
    Ohio 14, 23
Williamson, P.G. 199
Woo, Ju-Kang (Wu, JuKang) 3, 34, 39
World War I 22, 24
World War II 5, 6, 33, 43, 49, 52,
    218, 219
Wüist, E. 134
Würm glaciation 112, 127, 228

Young, C.C. vii, 33
Yuan-mou: incisors 74, 228, 233
Yunnan province 228

Zdansky, O. 22, 31, 32, 41, 64
Zinjanthropus 21